Maker Innovations Series

Jump start your path to discovery with the Apress Maker Innovations series! From the basics of electricity and components through to the most advanced options in robotics and Machine Learning, you'll forge a path to building ingenious hardware and controlling it with cutting-edge software. All while gaining new skills and experience with common toolsets you can take to new projects or even into a whole new career.

The Apress Maker Innovations series offers projects-based learning, while keeping theory and best processes front and center. So you get hands-on experience while also learning the terms of the trade and how entrepreneurs, inventors, and engineers think through creating and executing hardware projects. You can learn to design circuits, program AI, create IoT systems for your home or even city, and so much more!

Whether you're a beginning hobbyist or a seasoned entrepreneur working out of your basement or garage, you'll scale up your skillset to become a hardware design and engineering pro. And often using low-cost and open-source software such as the Raspberry Pi, Arduino, PIC microcontroller, and Robot Operating System (ROS). Programmers and software engineers have great opportunities to learn, too, as many projects and control environments are based in popular languages and operating systems, such as Python and Linux.

If you want to build a robot, set up a smart home, tackle assembling a weather-ready meteorology system, or create a brand-new circuit using breadboards and circuit design software, this series has all that and more! Written by creative and seasoned Makers, every book in the series tackles both tested and leading-edge approaches and technologies for bringing your visions and projects to life.

More information about this series at https://link.springer.com/bookseries/17311.

Industrial Metaverse Building Blocks

An Extended Reality (XR) Model-Based Approach

Alessandro Migliaccio
Giovanni Iannone
Mark Sage

Apress®

Industrial Metaverse Building Blocks: An Extended Reality (XR)
Model-Based Approach

Alessandro Migliaccio
Colomiers, France

Giovanni Iannone
Roma, Italy

Mark Sage
Bristol, UK

ISBN-13 (pbk): 979-8-8688-1171-5 ISBN-13 (electronic): 979-8-8688-1172-2
https://doi.org/10.1007/979-8-8688-1172-2

Copyright © 2025 by Alessandro Migliaccio, Giovanni Iannone and Mark Sage

This work is subject to copyright. All rights are reserved by the Publisher, whether the whole or part of the material is concerned, specifically the rights of translation, reprinting, reuse of illustrations, recitation, broadcasting, reproduction on microfilms or in any other physical way, and transmission or information storage and retrieval, electronic adaptation, computer software, or by similar or dissimilar methodology now known or hereafter developed.

Trademarked names, logos, and images may appear in this book. Rather than use a trademark symbol with every occurrence of a trademarked name, logo, or image we use the names, logos, and images only in an editorial fashion and to the benefit of the trademark owner, with no intention of infringement of the trademark.

The use in this publication of trade names, trademarks, service marks, and similar terms, even if they are not identified as such, is not to be taken as an expression of opinion as to whether or not they are subject to proprietary rights.

While the advice and information in this book are believed to be true and accurate at the date of publication, neither the authors nor the editors nor the publisher can accept any legal responsibility for any errors or omissions that may be made. The publisher makes no warranty, express or implied, with respect to the material contained herein.

Managing Director, Apress Media LLC: Welmoed Spahr
Acquisitions Editor: Miriam Haidara
Development Editor: James Markham
Project Manager: Jessica Vakili

Cover designed by eStudioCalamar

Distributed to the book trade worldwide by Springer Science+Business Media New York, 1 New York Plaza, Suite 4600, New York, NY 10004-1562, USA. Phone 1-800-SPRINGER, fax (201) 348-4505, e-mail orders-ny@springer-sbm.com, or visit www.springeronline.com. Apress Media, LLC is a California LLC and the sole member (owner) is Springer Science + Business Media Finance Inc (SSBM Finance Inc). SSBM Finance Inc is a **Delaware** corporation.

For information on translations, please e-mail booktranslations@springernature.com; for reprint, paperback, or audio rights, please e-mail bookpermissions@springernature.com.

Apress titles may be purchased in bulk for academic, corporate, or promotional use. eBook versions and licenses are also available for most titles. For more information, reference our Print and eBook Bulk Sales web page at http://www.apress.com/bulk-sales.

Any source code or other supplementary material referenced by the author in this book is available to readers on GitHub. For more detailed information, please visit https://www.apress.com/gp/services/source-code.

If disposing of this product, please recycle the paper

Table of Contents

About the Authors xi

Acknowledgments xiii

Introduction xv

Chapter 1: Awareness 1
   Why Extended Reality (XR)? 2
   Is It the Right Time to Talk About an Industrial Metaverse? 4
   Ontology: AR, VR, MR, XR, and Metaverse 6
      Engineering 6
      Virtual Reality 6
      Augmented Reality 7
      Mixed Reality 7
      Extended Reality 7
      However-Reality 8
      Are Three Dimensions Enough to Establish a Metaverse? 8
      What About the Multiverse? 9
      XR Gadgetries 11
      Training, Ergonomics, and Health 13
   Intersection with Artificial Intelligence (AI) 15
      Some Issues Linked to XR Technology 15
   XR: A Systems Engineer Experience 17
   Takeouts 19
   Questions 20

TABLE OF CONTENTS

Chapter 2: A Strong Modeling Foundation 21
Integrating Modeling, Simulation, and Uncertainty in Complex Systems 22
System Definition 24
MBSE Definitions 25
Test and Simulation 29
MBSE Advantages 31
Software Uncertainty 33
Monte Carlo Simulation 36
MBSE Applications 41
A Small Practical Example – Overview 43
A Small Practical Example – Simulation 49
Modeling Languages 54
Takeouts 56
Questions 60

Chapter 3: XR Life Cycle 61
The Life Cycle of Technology 62
System Life Cycle and XR 63
XR Applications in a System Life Cycle 65
Verification and Validation of XR 69
    Verification 69
    Validation 71
Digital Continuity 76
Takeouts 77
Questions 80

Chapter 4: Convergence 81
Convergence Primer 82
The Importance of Continuity and Life Cycle Mindset 85

TABLE OF CONTENTS

Emerging Disciplines 88
Current and Speculative Professional Roles 89
    Global or Local MBSE Architect 89
    MBSE Engineer with XR Expertise 90
    XR/Immersive Producer 90
    Virtual or Digital Engineer 90
    MBSE-XR Champion 90
    Narrative Designer 92
Impact of XR-MBSE Convergence on Industrial Activities 92
    Group Alpha – Immersive Mock-ups and Digital Twins 95
    Group Beta – Collaborative Design Reviews and Co-simulations 101
    Group Gamma – UML Diagrams and Functional Architecture 107
    Group Delta – Training and Field XR 109
A Proposed Metaverse MBSE Framework 112
A Brief Note on MBSE4XR 115
    The Metaverse As System of Interest 115
    XR Devices As System of Interest 116
    XR Applications As System of Interest 117
Takeouts 118
Questions 122

Chapter 5: Key Use Cases for XR 125
VR Applications in Your Work 126
Remote Assistance 129
Training 130
    Survey Conducted by AiShed 133
Guidance 139
Assembly 140

TABLE OF CONTENTS

Collaboration 142
Navigation 145
Maintenance 149
Realistic Experiences 151
Takeouts 152
Questions 153

Chapter 6: Deployment 155

Deploying in the Metaverse 156
    Examples from Industry 157
What the Metaverse Is and Is Not About 158
How to Harness the Potential of MBSE 162
Create a Community of XR Champions Within Your Organization 163
Encourage Intrapreneurship 165
How to Realize the Full Potential of XR 167
Return On Investment (ROI) 171
A Proposed Deployment Process 172
Do Not Reinvent the Virtual Wheel! 176
Cybersickness 177
    Recommendations for Safe XR Demonstration Sessions 179
Preparation of the XR Experience 180
Cybersecurity 181
Dealing with Unions and Work Councils 185
Takeouts 186
Questions 187

TABLE OF CONTENTS

Chapter 7: Learning from Experience 189
Interview with Brian Laughlin, PhD, TF (Boeing) 190
- The Evolution of XR in Aerospace 191
- Breakthrough Projects 191
- The Power of XR in Industrial Applications 192
- The Industrial Metaverse: Bridging the Experience Gap 193
- Integrating XR with MBSE 193
- Challenges and Best Practices for XR Implementation 194
- Conclusion: The Future of the Industrial Metaverse 194
- Interview with Ryan Wheeler (Collins Aerospace) 195
- Interview with Paul Davies (Boeing) 198
- Interview with Francis Vu (Immersified) 200

Procedural Training 202
Interview with Cedric Chane Ching (Aptero) 206
- Interview with Alexandre Mao (Torrus VR) 209

Appendix A: Artificial intelligence in the Metaverse 215

Appendix B: Sources 225

Glossary 229

Index 237

About the Authors

Alessandro Migliaccio, CEng, CSEP, is a graduate of Space Systems Engineering from Delft University of Technology and currently serves as a Business Product Leader at Airbus Operations. With expertise in mixed reality technology and aspirations as a futurist, he brings over ten years of experience in the aeronautical industry across multiple countries. Alessandro has spearheaded data analytics projects focused on optimizing and tailoring aircraft maintenance programs. As a Chartered Engineer recognized by the Royal Aeronautical Society, he is dedicated to enhancing team dynamics through innovative paradigms that emphasize professional fulfillment, talent nurturing, and the democratization of niche technologies. In pursuit of these goals, he founded AiShed, an association dedicated to outreach and research, driven by voluntary talents.

Giovanni Iannone holds a degree in Mechanical Engineering for Design and Manufacturing from Università degli Studi di Napoli in Italy and a master's degree in Systems Engineering from MS&T (USA). He is an expert in aeronautical structures and continuous airworthiness of large aircraft, bringing over ten years of experience as a subject matter expert. His work exhibits a

ABOUT THE AUTHORS

keen analytical interest in decision-making processes through various mathematical methodologies. As a long-standing member of INCOSE, he has contributed to the field by presenting on decision-making algorithms in the sports industry at ASEC2014.

Mark Sage is the CEO of Sagey11 Ltd, a consultancy company helping startups, scaleups, and large companies develop strategies and go-to-market plans to benefit from Enterprise Augmented Reality (AR) and Artificial Intelligence (AI) technologies. An MBA-qualified entrepreneur, Mark is also the Executive Director of the Augmented Reality Enterprise Alliance (AREA), the only global, membership-funded non-profit alliance dedicated to helping accelerate the adoption of Enterprise Augmented Reality (AR), as well as a subject matter expert for AR at MIT Horizons, the Digital Engineering Technology & Innovation (DETI) project, and a board advisor to an Australian AI startup. He enjoys mentoring startups and new CEOs and guest lecturing at various universities across the world on the benefits and how to successfully deploy enterprise AR solutions.

Acknowledgments

This book results from nearly three years of work and is the product of a decade-long professional partnership based on a continuous exchange of ideas. As is often the case, very few pieces of work come from individual minds. These usually grow into a metaverse of friends and supporters through late-night conversations and amiable chats by the coffee machine. Eventually, the activity became too large to require a structured approach. Therefore, we launched AiShed, a think tank headquartered in France but with a wide reach devoted to outreach and fuelling a community of like-minded people interested in world-changing technologies.

Figure 1. *Companies who contributed to the book*

Firstly, we must thank our families, friends, and colleagues, the interviewees: Brian Laughlin, Ryan Wheeler, Paul Davies, Francis Vu, Cedric Chane Ching, and Alexandre Mao, featured in Chapter 7, and their respective companies (some of them are partners of AiShed, in Figure 1)

ACKNOWLEDGMENTS

for spending their valuable time to share their hands-on experience with our readers. A final thanks to our colleagues Apoorv Somkuwar, Huanran Zhang, Samuel Martinez, and Michael Round for the valuable feedback on Virtual Reality training applications.

Thank you to all of you.

Introduction

Building Blocks of the XR Metaverse

Before we start our journey, one recommendation: put down the book, go to the nearest VR arcade, and try some of the experiences; appreciation for the technology comes with firsthand experience.

Neil Stephenson first introduced the term Metaverse in his science fiction novel *Snow Crash*, published in 1992. In this somewhat psychedelic novel, it is told for the first time that, through advanced technology, people can enter a virtual world by interacting with each other and engaging in various activities. The text's narrative brought ahead of its time, predicting with disturbing clarity some of the symbols of our time.

Although the Metaverse described by Stephenson takes on a controversial meaning by placing it in a dystopian context in which people are almost forced into a digital life to escape reality, we believe that its depiction has had a significant impact on popular culture and has influenced discussions about virtual reality and its potential applications. In particular, by transferring the concept of the Metaverse to the industrial environment, it is possible to imagine an immersive digital space, or a network of virtual and three-dimensional spaces, that could enable continuous interactions.

The most common thinking of industry organizations operating outside the perimeter of immersive technologies is to consider the Metaverse as something relegated only to video games or technology enthusiasts. While the application of virtual reality technologies in entertainment is a fascinating subject that could fill an entire book, it falls outside the scope of this one.

INTRODUCTION

This text does not pretend to convince the reader that virtual reality is a perfect substitute for the real world, but it does want to propose some thoughts that should move the consciousness of those who want to represent real experiences in a virtual world.

The concept of the metaverse has garnered significant attention in recent years, particularly in the realms of gaming and social media. However, the potential applications of this virtual and interconnected world extend far beyond industries. They can explore and revolutionize their operations by such virtual space offering the opportunity to digitize and optimize their processes, collaborate with partners and stakeholders, and unlock new levels of efficiency and innovation.

From manufacturing and logistics to healthcare, the industrial metaverse holds the promise of transforming how industries operate and interact with their surroundings. By leveraging virtual reality, augmented reality, and other immersive technologies, we can create digital twins of their physical assets, simulate complex scenarios, and experiment with new ideas in a safe and controlled environment.

In this rapidly evolving landscape, companies that embrace the industrial metaverse stand to gain a competitive edge by streamlining their operations, fostering collaboration, and staying ahead of the curve in a rapidly changing world. The industrial metaverse is not just the future – it's the present, and the time to explore its possibilities is now.

Verifying the potential of new technologies, introduced in the current industrial scenario that is not yet ready to be a metaverse concept, is a crucial step because all metaverse impacts shall be evaluated. In this text, all potential ways to verify and validate the effectiveness of these technologies will be described. Therefore, by verifying new technologies through these methods, businesses can make informed decisions about adopting the industrial metaverse and unlock its full potential for transforming industries.

It's worth underlining that virtual representations do not have to be the same as the real world, but they must be consistent and compliant with the rules used to observe, measure, and analyze reality.

INTRODUCTION

What does reality mean? What should we mean when we say real world? What does the expression "real world" mean? Real world and virtual reality are two distinct concepts that raise philosophical questions about the nature of existence, perception, and truth. According to a pragmatic approach, reality, like existence, is accessible when we refer to criteria of verification. We can always show that any assumption and definition of reality can be accurate based on an empirical, scientific, and physical approach. Therefore, reality is the tangible, physical world that exists independently of human perception (Figure 2). It encompasses everything that can be experienced through the senses and is governed by natural laws and objective truths. On the other hand, virtual reality is a constructed, digital environment that simulates aspects of reality but is fundamentally distinct from it. A "virtual reality" can mimic elements of reality, such as sensory experiences or spatial relationships; it is subject to human manipulation and design. Finally, it is a subject of simulated representation where the human perception and interpretation draw the boundaries between what is real and what is artificially constructed.

Figure 2. *Perception and reality*

xvii

INTRODUCTION

In conclusion, to accept that virtual reality is a perfect representation of the natural world is to admit that there are many digital universes, metaverses, in which the reality shown is assumed to be accurate within that world and certainly not outside of it.

Metaverses must be treated as different universes, and one must be careful not to confuse the planes in which they are employed in the predicate of existence. In short, between empirical truths and mathematical truths, commonsense knowledge and knowledge, so-called scientific knowledge, specific languages, and a consistent definition of reality must be ascribed.

Suppose we are operating in an airport and can peer from a distance at an object flying. From our experience, we can say that that flying object is an airplane, so our interpretation of reality defines that the flying object is an airplane. When the object approaches or is picked up by radar, it is confirmed to be a helicopter because its radios have transmitted an identifying signal that associates it with a helicopter. In this case, we can have two realities. One reality is subjective, related to perception and the other scientific, exact reality that a measurement of radio frequencies has defined. In this case, there is pragmatic evidence that refers to different universes. On the one hand, it is the perception and experience of observing an object from a distance, and on the other, it is the scientific measurement and analysis. These experiences define one universe in a perceptual sense, and another relates to scientific knowledge that requires one to read the data transmitted by radar.

What is the accepted reality? In some cases, it might be enough to identify an object that flies; in others, we need to specify whether it is a helicopter or an airplane. In both cases, we must set verification criteria to pursue different ascertainments. At this point, both realities can be true, but more importantly, they must be consistent and appropriate for the purposes under which reality is to be represented.

INTRODUCTION

According to neorealism, we can accept that reality accurately establishes that the object in flight is a helicopter rather than an airplane by an adequate degree of truth, assuming relative importance. Why? Because it depends on the point of view, or rather when that representation of reality has greater practical effectiveness.

Can we also consider "imagination" as a method of representing reality? Well, we can cultivate it if it suits us, because it is a mode of expression that might find some purpose and reveal its practical efficiency, in a pragmatic sense.

Ultimately, the representation of reality through the virtual world must be as effective as possible and true so that all the objects represented are closer to the truth and to a logic that best reconstructs reality through formulas, measurements, and empirical data.

With the advent of new technology, the Metaverse has brought about a remarkable convergence of cutting-edge reality that can be explored by use of 3D technologies, artificial intelligence, and blockchain. These advancements were already making significant progress individually, but the Metaverse has provided a platform for their seamless integration and collaboration.

In the industrial realm, the Metaverse has paved the way for a single source of truth – a centralized data lake that serves as a repository for all relevant information. This data lake enables real-time access to critical data and ensures all stakeholders work with accurate and up-to-date information. This unified approach allows businesses to achieve digital continuity across various processes and systems.

One of the key benefits of the Metaverse is its ability to facilitate model-based simulations. By leveraging 3D technologies and artificial intelligence, industries can create virtual representations of physical systems or processes. These simulations allow for accurate testing, optimization, and analysis before implementing changes in reality. This saves time and resources and minimizes risks associated with traditional trial-and-error methods.

INTRODUCTION

Furthermore, integrating extended realities within the Metaverse opens up new possibilities for immersive experiences and enhanced collaboration. Virtual Reality (VR) and Augmented Reality (AR) technologies enable users to interact with digital models in a more intuitive manner. Teams can collaborate remotely in shared virtual spaces, breaking down geographical barriers while fostering creativity and innovation.

Another interesting point is how can we ensure a level of security in our data management. The Blockchain technology is introduced in the Metaverse framework ensuring transparency, security, and trust in data transactions. Its decentralized nature eliminates intermediaries while providing an immutable record of every interaction or transaction within the ecosystem (see interview of Alexandre Mao in Chapter 7).

In conclusion, the advent of the Metaverse has propelled advancements in 3D technologies, artificial intelligence, extended realities, and blockchain to new heights. The integration of these technologies allows industries to harness their collective power for improved collaboration, streamlined operations through a single source of truth concept such as data lakes or digital continuity models-based simulations, and enhanced experiences within a secure and transparent environment. The future possibilities of the Metaverse are boundless, offering endless opportunities for innovation and transformation across various sectors. A virtual interaction would overcome all those physical and economic obstacles, present in the real world, that could compromise the successful implementation of any project.

The SE Approach

To date, the Systems Engineering (SE) approach provides a valuable tool to ensure that all phases of a project work together efficiently. One of the techniques used by the SE approach is definitely system modeling (refer to Chapter 2 for further discussion). Modeling is a way of representing a system and understanding how it works. In addition, proper modeling succeeds in identifying potential problems and testing possible solutions. In general, to create a model of a system, observation plays a crucial role, but to date tools that can define it are not so effective and thus its explanation defers to considerable mental effort.

Therefore, the industrial Metaverse would allow users to interact with virtual models, more or less detailed, through a multisensory experience, which in turn would ensure that the virtual model is consistent with the real thing, while also avoiding a considerable deployment of mental production of our thinking.

The content of the book addresses how XR can be revolved around the life cycle of a generic system and can support the main activities in different phases of system development.

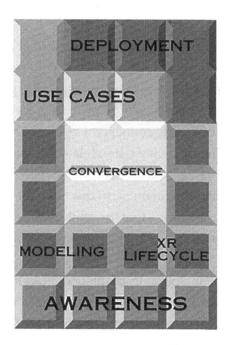

Figure 3. Building blocks of the Industrial Metaverse

INTRODUCTION

The text is structured in six building blocks/chapters (Figure 3) that give the reader the process needed to understand how to create, support the industry activities using an immersive technology. In such a perspective, interesting expert interviews are collected in the Appendix, in which the reader can find several players' experiences that made their primary mission to deploy Extended Reality in their organization. Despite this, this book is not meant to compete with the research of hundreds of experts all over the world working to advance the frontiers of Extended Reality (XR), the Metaverse, and Model-Based Systems Engineering (MBSE); its aim is simply to help the reader to put into practice the theoretical framework presented in the next pages through some useful case studies.

The journey to understanding the industry Metaverse is a fascinating one; although it can be perceived as arduous to the inexperienced reader, we believe that the first steps for a factual metaverse implementation shall be done.

The industrial Metaverse can be introduced either at the beginning of a project or in projects already developed; in either case, we need to find the right cue from the Systems Engineering (SE) approach. In fact, we recommend a basic knowledge of SE principles, but over and above, we hope that the text can easily connect the reader to bullet points of a systematic, holistic, and modeling vision (also check interview with Dr. Laughlin in Chapter 7).

Finally, let us reach an agreement: in this book, we will examine the basics of the topic, assuming that the reader will be proactive in utilizing the resources available on the Internet or in literature to further understanding.

You can read the book chapter by chapter or by picking a specific chapter/block of interest. This book is enriched with examples from classic literature and examples from industry to help you grasp the most difficult notions.

We have also created a website we invite you to visit, hoping to intrigue your curiosity: www.ai-shed.com.

We hope you find this an entertaining and useful read!

CHAPTER 1

Awareness

> *VR and AR can seem very new. The question is, "When to start investing, when to start building?". If you've ever been surfing, you know that you need to start paddling well before the wave starts to crest if you're going to catch the wave. You watch the water, you look at the ocean, you look for raises in energy, everything lining up.*
>
> *I spend my days reading the waters, seeing if everything is going to line up, seeing when that energy is going to start to build. Here, with VR and AR, I will just say that now would be a really good time to start paddling.*
>
> —Clay Bavor, VP of Virtual Reality, Google
>
> SID Week, Los Angeles, June 2017

Chapter 1 introduces the concepts of Virtual Reality (VR) and Augmented Reality (AR), emphasizing the importance of investing in these technologies early on. The chapter discusses the potential of Extended Reality (XR) as the fourth computing platform revolution, following PCs, the Internet, and mobile phones. It highlights the growing interest in XR due to its potential for profit and its ability to transform ideas into accessible and sellable products.

The chapter also explores the concept of the metaverse, a virtual reality space where people can interact with digital content. It discusses the rise of Bitcoin as an early example of an industrial metaverse and the potential

CHAPTER 1   AWARENESS

for further development in this area. Additionally, the chapter covers the importance of understanding key aspects of XR technology, including low-code and no-code solutions, and the role of tools like Unity, ZapWorks, and Blippar in making XR accessible to a wider audience.

Overall, this chapter sets the foundation for understanding XR technology and its potential impact on various industries, encouraging readers to start exploring and investing in this emerging field.

Why Extended Reality (XR)?

The answer to this question is not to be found in metaphysics but in something largely more prosaic. High-tech is a potential source of great profit, and for this reason, there are more and more companies interested in transforming the idea into a real thing, easily accessible and sellable.

You are likely reading this book because you have perceived a future in immersive technology, and you want to be ready to have a piece of the action, either as a manager, business person, developer, producer, designer, or simply as a gamer looking for some escapism at the end of a long day of work.

Do you believe it to be an easy goal to achieve? The main suggestion is to immerse yourself in a particular Systems Engineering discipline, which can help you to understand why we have to talk about models. By utilizing models as representations of reality, Model Based Systems Engineering (MBSE) allows for the creation of virtual environments that closely mimic real-world scenarios.

During our trip to discover the world of immersive (or metaverse) tech, we will try to define the ideas behind this future vision. The development of technologies has led us to transform different views of reality into something usable for a particular scope. To define a place where you find the metaverse or "a virtual world," we have to cross the threshold that separates the "real" from "virtual" reality and define the different degrees of separation between them.

CHAPTER 1 AWARENESS

XR is expected to become the fourth computing platform revolution after PCs, the Internet, and mobile phones (Figure 1-1). Enabling the implementation of disruptive use cases that other technologies simply cannot, it is already beginning to impact society across the board, changing the way people consume and interact with information. However, many are wondering if XR has the potential to move from POC (Proof of Concept) to an industrial scale.

In this chapter, we set the foundation by providing general awareness of a number of key aspects of XR technology; by Chapter 6, all the building blocks should be in place to give an answer to the question about the scalability of XR. Clearly, the more cash flows into the sector, the more tools are created to help new people enter the fold. Even if you have the coding capabilities of a wooden spoon, you must be aware of low-code applications that anyone can pick up and make basic applications after a little self-teaching. Tools like Unity, ZapWorks, and Blippar offer a range of ways for developers to get started, with a wealth of materials and courses to kick-start the process. With that ease comes light barriers of entry for anyone to access.

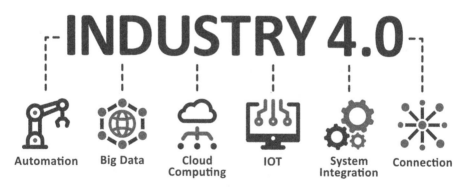

Figure 1-1. *Industry 4.0*

Professionals like you are looking into such low-code solutions, and companies have stepped forward to plug the gap in the market with some (profitable) courses. Meanwhile, other platforms provide no-code

3

solutions that allow people to create experiences by stringing together templates. As said in the introduction, this book aims to give the first guidance on moving from DIY XR to a full-scale Industrial Metaverse.

Is It the Right Time to Talk About an Industrial Metaverse?

The concept of a metaverse (Figure 1-2), a virtual reality space where people can interact and engage with digital content, is gaining traction in various industries. One of the first examples of an industrial metaverse can be seen in the framework of virtual money, specifically through the phenomenon known as the Bitcoin fad.

Bitcoin, a decentralized digital currency, has captured the attention and imagination of individuals worldwide. It represents a significant step toward creating a virtual monetary system that operates outside traditional financial institutions. Built on blockchain technology, Bitcoin can be seen as an early manifestation of what could eventually evolve into a fully-fledged metaverse.

The Bitcoin fad highlights how virtual currencies have garnered immense popularity and sparked widespread interest. It demonstrates how people are willing to invest time, resources, and trust in digital assets that exist solely within the realm of the virtual world. This enthusiasm for Bitcoin serves as evidence that individuals are embracing the idea of participating in an industrial metaverse where transactions occur digitally and independently from physical constraints.

CHAPTER 1 AWARENESS

Figure 1-2. *Abstract illustration of the "Industrial Metaverse"*

As we witness this first example of an industrial metaverse through the rise of Bitcoin, it becomes clear that virtual experiences are becoming increasingly intertwined with our daily lives. The potential for further development and expansion within this realm is vast, offering exciting opportunities for innovation across industries beyond finance.

In conclusion, the advent and widespread adoption of Bitcoin exemplify one of the earliest instances where we see elements of an industrial metaverse taking shape. As technology continues to advance and more industries explore virtual spaces, it is likely that we will witness further examples that push boundaries and redefine our understanding of what is possible within this evolving digital landscape.

CHAPTER 1   AWARENESS

Ontology: AR, VR, MR, XR, and Metaverse

Before we move any further in this technological roadshow, it is important to give meaning to some of the keywords used in the book from this point onward, at least what they mean in an engineering context. Furthermore, this book will refrain as much as possible from citing specific commercially available virtual reality hardware so that the reader shall be bound less by the limitations of current technology and focus more on the problem rather than the solution.

Engineering

An engineer is a person who has a degree from an engineering school. If you believe that, I will endeavor to expand your views somewhat. It may seem pointless to define what the authors mean by engineering, but it is essential to have a moment of alignment on this key point. An engineer is someone who will use ingenuity to create something, or in our specific domain, someone who would use technology to create value, to add to the substance of the world, and to benefit mankind. Similarly, systems engineering is a way of thinking, not a badge to obtain by passing an exam. As said in the introduction, everyone asking themselves the right questions is not just an engineer but a "systems engineer."

Virtual Reality

With virtual reality, we mean the generation of a simulated reality within which the senses of sight and hearing are isolated from the external (real) world and excited by synthetic stimuli (images and sounds) created with a computer. We talk about "immersive virtual reality" when all human senses are excited by artificially created stimuli: in this case, stimuli are added to images and sounds that excite the senses of touch, heat perception, proprioception, and – theoretically – also those of smell and taste.

Augmented Reality

With augmented reality, we mean the enrichment of human sensory perception through information (e.g., Hololens Remote Assistance), generally manipulated and conveyed electronically, which would not be perceptible with the five senses concurrently.

Mixed Reality

Mixed reality provides a regular choice between the two technologies presented. It is the seamless blending of reality that we can see directly through the human eye and different augmentations provided by digital information, often overlaid with reality.

Extended Reality

XR is what you make of it. More specifically, it is an umbrella term used to cover the overall of different domains. Since the perimeter of AR, VR, and MR is continuously evolving and is likely to be fluid for at least the next decade, it is rather pointless to spend any meaningful amount of time trying to define its boundaries exactly (Figure 1-3).

CHAPTER 1   AWARENESS

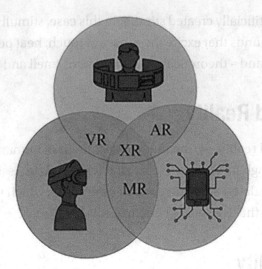

Figure 1-3. *The overlapping domains of XR*

However-Reality

At the time of writing, immersive technology is still fluid, and a myriad of buzzwords are conceived every single year by avid marketers to push a variety of gimmicks while new trends are born every month. The sole attitude recommended at this stage is to look beyond the surface to the real opportunities provided by the tech, like a producer scouting the landscape for opportunities to create a new piece of art, regardless of what is considered trendy.

Are Three Dimensions Enough to Establish a Metaverse?

In the vast and immersive world of the metaverse, time takes on a whole new dimension. As we navigate through virtual realms and explore multiple dimensions, the concept of time becomes both fascinating and complex.

Unlike our physical world, where time is measured linearly, the metaverse offers a unique perspective on this fourth dimension. Time in the metaverse can be manipulated and experienced in ways that defy traditional constraints. It allows for a fluidity that transcends our understanding of time as we know it.

In this digital realm, users can seamlessly navigate between different spaces and interact with others across vast distances instantaneously. Time becomes malleable, allowing for simultaneous experiences and interactions across various virtual environments.

Furthermore, within each dimension of the metaverse, time may operate differently based on its own rules and parameters. Some virtual worlds may have accelerated or slowed-down timelines, while others may exist outside conventional notions of past, present, and future.

The evolution of technology continues to shape our perception of time in the metaverse. As advancements in virtual reality (VR) and augmented reality (AR) propel us deeper into these immersive digital landscapes, our understanding of temporal boundaries expands alongside it.

As we venture further into this dynamic realm where possibilities are limitless, it becomes clear that time within the metaverse is not just a linear progression but an intricate tapestry woven through multiple dimensions. It challenges us to redefine our understanding of temporality and embrace the boundless nature of this thrilling digital frontier.

What About the Multiverse?

The concept of the industrial metaverse is evolving, and it's becoming clear that a single, unified metaverse may not be the reality (Figure 1-4). Instead, we're seeing the emergence of an industrial multiverse, where separate, siloed metaverses exist within different companies and industries.

CHAPTER 1   AWARENESS

This fragmentation raises concerns about industrial espionage, as data lakes and other assets within each metaverse may be vulnerable to theft or unauthorized access. Export control regulations and a lack of communication between metaverses could further complicate this landscape.

Figure 1-4. *Abstract illustration of the "Industrial Multiverse"*

However, as the industrial multiverse takes shape, we may see links established between these separate "golden islands," allowing for controlled data sharing and collaboration. Each company's metaverse

CHAPTER 1   AWARENESS

could become its own self-contained world, with carefully curated access and interactions. This interconnectivity could lead to new forms of business partnerships and collaborative innovations. Companies might establish "trade routes" between their metaverses, allowing for secure exchange of digital assets, virtual products, or even virtual workforce collaborations. These connections could be governed by smart contracts and blockchain technology, ensuring transparency and security in cross-metaverse interactions. Moreover, standardization efforts may emerge to facilitate seamless navigation between different industrial metaverses. This could include the development of universal avatars, interoperable digital currencies, or shared protocols for data exchange. As these standards evolve, we may see the rise of metaverse consultants and integration specialists who help companies navigate the complexities of inter-metaverse operations.

The future of the industrial metaverse is likely to be a complex, multi-faceted ecosystem, where companies navigate the challenges and opportunities of this new digital frontier.

XR Gadgetries

When it comes to XR (Extended Reality) hardware, attempting to classify them into specific categories can be a challenging and subjective task. However, for the sake of amusement, here are some common classifications that are often used:

- HMD (Head-Mounted Display): This refers to devices that are worn on the head, like a pair of goggles or glasses. HMDs can provide different degrees of immersion and interactivity depending on their features.

CHAPTER 1 AWARENESS

- Degrees of Freedom: This classification refers to the level of movement and tracking capability offered by XR hardware. It can range from stationary devices that only track head movements to more advanced systems with full six degrees of freedom (6DoF) tracking for both head and hand movements.

- Stationary/Desktop Scale: These XR experiences are limited to a fixed location or desktop environment. Users typically interact with virtual content using input devices such as controllers or keyboards.

- Room Scale: Room-scale XR allows users to move around within a designated physical space while being tracked by sensors or cameras. This enables more immersive experiences where users can physically walk around and interact with virtual objects.

- Marker-based: Some XR systems rely on markers or fiducial markers placed in the environment for tracking purposes. These markers help the system understand the user's position and orientation in relation to virtual content.

- Markerless/Emi-immersive: In contrast to marker-based systems, markerless XR does not require any external markers for tracking. Instead, it uses computer vision algorithms and sensors built into the hardware to track user movements.

- Fully Immersive: This classification refers to XR experiences that provide a high level of immersion by combining realistic visuals, spatial audio, haptic feedback, and advanced interaction capabilities.

- Tethered: Tethered XR devices are connected via cables or wires to a separate processing unit such as a computer or gaming console. This allows for more powerful graphics and processing capabilities but limits the user's mobility.

- Stand-alone: Stand-alone XR devices are self-contained and do not require any external processing units or cables. They have built-in processors, storage, and connectivity options, providing users with freedom of movement.

- 3D Glasses: This term is often used to refer to older forms of stereoscopic glasses that provide a basic 3D effect by separating the visual information for each eye.

It is important to note that these classifications are not exhaustive, and XR hardware continues to evolve rapidly, blurring traditional boundaries between categories.

Training, Ergonomics, and Health

The metaverse, a virtual reality space, offers numerous opportunities for learning and development. One of the key use cases for the metaverse is immersive learning, which leverages virtual spaces to simulate "real" work situations. By immersing employees in these virtual environments, learners can engage in hands-on experiences that enhance their understanding and skills.

Immersive learning with XR in the metaverse incorporates proven behavioral learning best practices. Learners are actively involved in realistic scenarios, allowing them to practice and apply their knowledge in a safe yet lifelike setting. This active participation enhances learner engagement by providing an interactive and dynamic learning experience.

CHAPTER 1   AWARENESS

Furthermore, immersive learning with XR in the metaverse improves readiness by preparing learners for real-world challenges they may encounter on the job. The ability to navigate through virtual workspaces and interact with virtual colleagues or customers helps build confidence and familiarity with different scenarios.

Another advantage of immersive learning in the metaverse is its impact on knowledge retention. Studies have shown that experiential learning leads to higher retention rates compared to traditional classroom-based instruction. By engaging multiple senses and creating memorable experiences, immersive learning promotes long-term retention of information.

Fortune 500 companies are realizing these benefits at scale with immersive platforms and skill-building in situational awareness, operational procedures, security, customer service, and soft skills. Despite the advantages, the industrialization of VR has hit some limits, some of which can be eventually bypassed with clever technological advancements and societal changes, but not all.

In Chapter 6, we will delve into the various aspects of XR testing and implementation, with a particular focus on the role of ergonomics in limiting the widespread use of VR hardware and certain AR devices. One key factor that hinders the broader adoption of these technologies is the issue of safety and health considerations.

Ergonomics plays a crucial role in ensuring user comfort and minimizing potential health risks associated with extended XR usage. The design and functionality of VR headsets, for example, need to prioritize factors such as weight distribution, adjustable straps, and proper ventilation to prevent discomfort or even injury during prolonged use. Moreover, XR sickness is another significant concern that arises from poorly designed or implemented XR experiences. This condition, similar to motion sickness, can manifest in symptoms like nausea, dizziness, and eye strain. By understanding the causes and implementing appropriate measures to mitigate XR sickness, developers can enhance user experience and encourage wider adoption.

In this chapter, we will explore various strategies employed by industry professionals to address safety concerns related to ergonomics and combat XR sickness. By prioritizing user well-being through ergonomic design principles and minimizing adverse effects like XR sickness, we can pave the way for a more accessible future for virtual reality (VR) and augmented reality (AR) technologies.

Intersection with Artificial Intelligence (AI)

Artificial Intelligence (AI) is important to the Metaverse as a driver of enterprise research in areas such as content analysis, self-supervised language processing, robot interaction, computer vision, and whole-body pose estimation.

AI can be used to deliver metaverse business applications in a number of ways. A subfield of AI, "AIOps," uses machine learning to help organizations manage their IT infrastructure. This will soon be applied to the Metaverse system. Additionally, AI-powered chatbots are becoming increasingly popular among businesses. AI bots with realistic avatars can be used for a variety of purposes in the metaverse, including sales, marketing, and customer support. The topic of integration of AI in the Metaverse is covered in Appendix.

Some Issues Linked to XR Technology

The future of XR (Extended Reality) technology is not likely to be dominated by a single solution but rather a combination of different modalities. This means that hybrid solutions incorporating various aspects of XR, such as VR, AR, and MR, will become the norm (Figure 1-5).

CHAPTER 1   AWARENESS

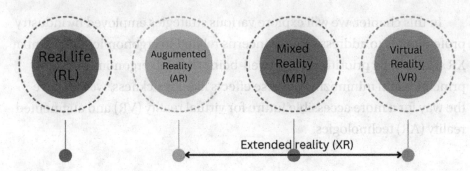

Figure 1-5. Diagram of the different "Reality" classification at time of writing

While each modality has its own unique capabilities and limitations, relying solely on one form of XR technology may lead to obsolescence and restrict the potential applications. By combining different modalities, we can overcome the limits of individual technologies and create more immersive and versatile experiences. For example, imagine Virtual Reality devices with AR features that allow users to overlay virtual objects onto their real-world environment. This could enhance training simulations or gaming experiences by providing additional contextual information.

Similarly, cars equipped with MR features could provide drivers with real-time information about their surroundings, such as navigation prompts or hazard warnings overlaid onto the windshield. This would improve safety while driving by reducing distractions.

Moreover, XR technology opens up possibilities beyond personal devices. We could envision spaceflights being experienced in holographic form within our living rooms, allowing us to follow missions in real time without leaving our homes.

In conclusion, the future of XR lies in embracing multimodal solutions that combine different technologies to overcome limitations and enhance user experiences. By doing so, we can unlock the full potential of XR and create immersive environments that were once only imaginable.

XR: A Systems Engineer Experience

In Chapter 2, we will delve into the concept of systems and explore its significance in the field of system engineering. However, for proactive systems engineers, it is hard to ignore the promises of XR as a key element of digital engineering. Digital engineering has revolutionized the way systems are designed and developed. It involves leveraging advanced technologies and tools to create model-based representations of complex systems. These models enable engineers to simulate, visualize, and analyze various aspects of a system before it is built physically. XR adds an extra dimension to digital engineering. By immersing users in virtual environments or overlaying digital information onto the real world, XR enhances collaboration, communication, and decision-making throughout the system development life cycle.

For proactive engineers who strive for efficiency and innovation, XR offers numerous benefits. It allows them to visualize system components in a realistic manner, identify potential design flaws or optimizations early on, conduct virtual testing and validation exercises, train personnel in simulated environments, and even provide interactive user experiences for stakeholders. By embracing XR as a key element of digital engineering practices, proactive engineers can unlock new possibilities for system design and development. They can enhance their ability to create robust and optimized solutions while reducing risks associated with physical prototyping or post-implementation modifications.

It could be argued that to incorporate a technology into a product and to avail of its full potential, an in-depth knowledge of the subject would be absolutely essential. The counterargument to that is that a systems engineer, oftentimes, plays the part of an integrator of a "system," or the producer of an immersive experience by observation.

CHAPTER 1   AWARENESS

Some amusing anecdotes can be drawn from personal experience, we shall recall that the authors are children of the 90s, so the first experience of immersion we have outside the ocean were, for the lucky ones, some bulky machines installed at amusement parks as temporary attractions (Figure 1-6).

Figure 1-6. *Virtuality system, Stockholm Museum of Technology*

The system was absolutely cumbersome and involved carrying a small TV on your face, resolution was extremely poor and experiences were limited to a few video games. Yet it was enough to spark the imagination and to leave the door open for further connection to the tech later on in life, the nostalgia factor if you will.

CHAPTER 1   AWARENESS

Now moving to 2015, after the failure of the ambitious Google Glass, some early Oculus devices hit the labs of engineering firms where they came in contact with some curious enterprising employees, eager to push the immersive technology to beat the competition. Some fortunate engineers have had the opportunity to try the Microsoft HoloLens, out of reach for most people, and from that point on, the words "industrialization" and "scaling" came up as a real possibility. Together with quantum computing, 3D printing, and many others, early VR took its place on the shelf of the promising niche technologies for the decade to come.

Takeouts

- Introduction to XR Technologies: The chapter introduces Virtual Reality (VR) and Augmented Reality (AR), emphasizing the importance of early investment in these technologies. It highlights XR (Extended Reality) as the fourth computing platform revolution, following PCs, the Internet, and mobile phones.

- The Metaverse and Bitcoin: The chapter explores the concept of the metaverse, a virtual reality space where people can interact with digital content. The chapter mentions the rise of Bitcoin as an early example of an industrial metaverse and the potential for further development in this area.

- Accessibility and Tools: The chapter covers the importance of understanding key aspects of XR technology, including low-code and no-code solutions. It mentions tools like Unity, ZapWorks, and Blippar that make XR accessible to a wider audience.

CHAPTER 1   AWARENESS

Questions

- Some say the future is what will happen in ten years for now. Can you imagine the future of XR?
- Name at least three key challenges of XR that will be addressed in the next ten years.
- In the chapter we talked about ergonomics, but what about safety, privacy, and cybersecurity? Search the web for those topics in connection to XR.

CHAPTER 2

A Strong Modeling Foundation

> ...the formalized application of modeling to support system requirements, design, analysis, verification and validation activities beginning in the conceptual design phase and continuing throughout development and later life cycle phases.
>
> —INCOSE SE Vision 2020

This chapter presents the interconnection between modeling, simulation, and uncertainty used for understanding and analyzing complex systems such as virtual reality. The integration of modeling, simulation, and uncertainty is essential for robust decision-making that represents the crucial feature of a complex system. By accounting for uncertainties during the modeling phase, one can create more resilient systems and develop strategies that are effective under a range of possible future scenarios. Sensitivity analysis, uncertainty quantification, and scenario-based simulations are common practices employed to better understand how uncertainties affect outcomes. Modeling and simulation serve as tools for exploring and analyzing complex systems, while effectively addressing uncertainty is crucial for making informed decisions based on these analyses.

CHAPTER 2   A STRONG MODELING FOUNDATION

Integrating Modeling, Simulation, and Uncertainty in Complex Systems

A philosophical overview has been given in the introduction, leading us to treat the real world through a virtual reality. When we wish to use XR technology for industrial purposes, it is necessary to first understand and define the real world and then focus on how to represent it in a virtual reality. In this chapter, we will try to explain how to represent reality through models, because in the XR framework, virtual objects and virtual environments are the pillars for creating immersive experiences. In one word, digital representation (see interview of Alexander Mao in Chapter 7).

A logical step, which we must strive to take, is to associate the observed reality with a generic system (complex, SoS, physical, etc.). Let us sit in a chair and observe everything around us. This is a matter of performing a small analysis of the reality around us, rather than observing our workplace or the trains and passengers flowing through a train station. We are observing a series of entities (in the sense that they formally belong to reality) that, while intuitively different from each other in form and substance, are mutually interconnected and interact with each other and the external environment.

Again, if modeling is nothing more than an abstraction of reality, and if reality is associated with a system, then we can represent the system through a model.

The construction of models, whether they are representative of the language or are just geometric representations, characterized by a particular logical-mathematical syntax, is initially determined by experience alone and then by a careful verification phase. These representations, in the form of models, obtained by systematic analysis of many classes of events, go back to concrete experiences. Therefore, common experience (common sense) would initially lead to the correct representation of an object in the virtual world; but we will not stop there!

CHAPTER 2 A STRONG MODELING FOUNDATION

Although we believe that the MBSE approach can be integrated into the XR technology design process, the use of digital models can also be emphasized in other phases, analysis, simulation, and maintenance of a generic system (Figure 2-1). This approach improves collaboration among teams by facilitating processes and activities at the multidisciplinary level because a complete understanding of the system to be designed is guaranteed.

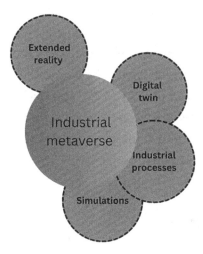

Figure 2-1. *Simplified diagram of the industrial metaverse*

In 1986, international professor and author B. Dijkstra introduced MBSE, which stands for Model-Based System Engineering. He proposed that software quality could be improved through a deterministic and constructive approach, made possible by the use of models. MBSE aims to improve system engineering by using models to help designers better understand and visualize complex systems. This approach has become increasingly popular in recent years and is now used in a wide range of industries, including defense, aerospace, and software development (see Interview of Dr. Laughlin in Chapter 7).

CHAPTER 2   A STRONG MODELING FOUNDATION

A model-based approach would allow keeping all relevant information in one place, as documents (drawings, test reports, etc.), requirements management package, abstractions, and even other models (Model of Models) converge in the proposed model. Understand well that if we needed to update documents more frequently rather than project requirements or/and the system itself, having a single model would allow us to work in tighter time frames.

In the introduction, we tried to define the degree of truth because of associating with the observed reality. Defying what is real from what is not real is not the scope of our investigation, especially when we use inappropriate methods which have to be involved in representative models of reality. Well, through the MBSE approach we can use a single system model that serves as a "source of truth" for all systems analysis tools. A system model can then interact with a virtual reality environment to analyze system performance.

System Definition

In the context of XR, the concept of systems plays a crucial role as it is often what is built, analyzed, or maintained through immersive technology. To fully understand this concept, it is important to define what a system actually is.

A system can be defined as a collection of interconnected elements that work together to achieve a common goal or purpose. These elements can include components, processes, people, and technologies that interact with each other in a coordinated manner.

In the context of XR, a system of interest refers to the specific focus or area that is being addressed or explored using immersive technology. It could be anything from a virtual training program for healthcare professionals to an augmented reality application for architectural design. By understanding and analyzing systems within XR, we can better comprehend how different elements come together to create immersive experiences and how they can be optimized for various purposes. This understanding allows us to build more effective and impactful applications within the realm of extended reality.

MBSE Definitions

The knowledge of a system depends on observation of it; in other words, the knowledge of a system is more limited when the observation is not accurate up to the detail. When a real system, observed up to a specific level of detail, is associated with a model, physical or mathematical, we must take into account that all information about the system state and behavior are observable and measurable. The "observability" of a real system requires that

- The observer chooses what to measure.
- The observer chooses a reference to use for the measurement.
- The observer chooses a measuring instrument.

Note Not all behaviors of a system can be observed directly but they can be known by inference due to cause-and-effect concatenation.

Creating a system model involves understanding the intricacies of its engineering and design. It requires us to simulate real-world scenarios and test the performance of our solutions before implementation. By doing so, we can identify potential issues or challenges that may arise during actual usage. Tangible experience and experimental tests are often sufficient to analyze a system and consequently make a finished product meet the

CHAPTER 2   A STRONG MODELING FOUNDATION

stakeholder's expectations. Spending plenty of resources to create a model that fits our system in all its complexity can appear as a great contraction when we think about the allowances necessary for computational activities of model creation, simulation and testing.

Certainly, a first answer is provided by the meaning of system engineering, that is, to program in a systematic way all those activities that characterize the simulation and reproduction of a model built for experimental purposes and, more generally, as a representation of a system designed to study the behavior of certain physical quantities, in order to optimize a project.

In many engineering disciplines, people choose to perform numerous calculations prior to the completion of the design phase so as to reduce the number of prototypes and tests required, but more importantly, it will enable the creation of objects or structures with an optimized design, all thanks to analyses that cost significantly less than "traditional" laboratory experimentation.

Therefore, it is enough to convince oneself that introducing a numerical simulation activity from the earliest stages of study is not an additional expense, but a saving on the overall project cost.

When discussing Model-Based Systems Engineering (MBSE), it is useful to introduce the concept of System of System (SoS), which is necessary to identify all the operational and functional interdependencies of all system components, in order to correctly allocate behavior and processes related to model development. Conversely, the modeling approach may be confusing and labile. A system, having the above-mentioned characteristics, is considered a SoS, in other words, the identification of all component independencies shall be easily defined, regardless of the complexity of the considered systems. Figure 2-1 shows a complementary diagram of what happens when a SoS is modeled, taking into account the behaviors and iterations with the environment of each individual element.

ISO/IEC/IEEE 21839 (ISO, 2019):
"System of Systems (SoS) — Set of systems or system elements that interact to provide a unique capability that none of the constituent systems can accomplish on its own. Note: Systems elements can be necessary to facilitate the interaction of the constituent systems in the system of systems."

The goal of model-based is to foster the development and optimization of increasingly complex, efficient, and reliable products, as well as to foster important pointers to a reality that is often referred to as "virtual."

At this point, it is worthwhile to understand whether the definition of a model, its simulation, and its virtual representation saves us time and costs during the product life cycle. Therefore, MBSE is an emerging approach in the field of SE and can be described as the formalized application of modeling principles, methods, languages, and tools to the entire life cycle of complex, interdisciplinary socio-technical systems. The simplified definition of MBSE, provided by Mellor as "... is simply the concept that we can build a model of a system that we can transform into a real system" helps us carry MBSE into the virtual reality where we need to manage complex systems or systems of systems.

The essence of MBSE is based on the application of models that can be applied to a system or even to systems that already exist in reality, in which boundaries have not yet been demarcated. These can relate both to technical aspects, such as interfaces, integration, and testing, and to managerial aspects, such as governance.

Therefore, the systems engineering approach focuses on the boundaries and interactions between seemingly independent and evolving systems.

The systems of an SoS are operated by initially independent entities that exchange information limited to their operational scope. Suppose we are at an airport as usual, in which both airplanes and air traffic controllers and all personnel dedicated to maintenance and customer service operate. Consider only the airplane system (complex system), it would appear to

CHAPTER 2 A STRONG MODELING FOUNDATION

be operated independently, but as soon as it encounters another airplane or needs to land it begins to exchange information with the other systems in the SoS (airport). Therefore, the behavior of the airport SoS is related to the behavior of each of its individual systems. Finally, knowing that each individual system has its own life cycle, it inevitably follows that the complexity of the SoS may increase due to the overlap of multiple life cycles. Based on the above, stakeholders may face problems of lack of a centralized authority that would have the purpose of managing them simultaneously.

For the MBSE application to be effective, the support of architecture models (System Functional Flows – System Architecture) starting from requirements (System Requirements Traceability) defined by stakeholders is necessary to obtain an intuitive representation of the analyzed system. Additionally, to better manage the complexity of the system, it is recommended to create optimized and combined processes (System and Organizational Process Flows) to save time and reduce costs.

Let us try to understand this last concept better. In general, in order to create a model-based system for any engineering system (system for configuration control, safety analysis system, software development system, etc.), one must consider both informal design models and requirements documents. Now, it is quite clear to understand that because of the informal models used as the basis of the analysis, the results are not always error-free, complete, and consistent. At this point, more accurate models must be sought from the system architecture. In model-based, in fact, the main development activities such as simulation, verification, testing, and code generation are based on a formal model of the system that should define the behavior of the system and allow it to automate parts of the analysis process, reducing costs and improving the quality of the processes.

Proper system architecture allows the creation of simplified and customized model views to isolate components of interest according to different needs. It is possible through simulation to validate requirements and verify the above architecture.

It is clear that the additional complexity due to multiple system life cycles, as well as the balancing of different requirements, carries over to the testing phase, and affects the performance of the various systems constituting the SoS.

Test and Simulation

When we talk about virtual reality or extended reality, we must not overlook that individual consciousness can influence experiences or observations in a subjective way. This can negatively affect the use of VR for engineering purposes. Individual bias could influence virtual representation, leading us to different levels of abstraction which may not be suitable in the technical field. The first question to ask ourselves is definitely: to what degree can we consider virtual activities a truthful representation of the activities carried out in the physical world? Therefore, a correct simulation can prevent our biased perception of reality from leading to wrong choices during activities required in the engineering framework. In engineering and in science, the purpose of simulation is to imitate and reproduce, as closely as possible, the behavior of a particular observed system. Simulation, consisting of one or more computational methods, is used to analyze and study a model representative of the system behavior. Analyzing a system involves numerous factors and variables, many of which can significantly increase the computational demands of our digital model. Without the use of specialized software to facilitate the ongoing search for results, this complexity can become overwhelming.

For the MBSE philosophy, a numerical simulation is very useful since it provides artificial but concrete models, in other words, some models can reproduce reality an infinite number of times.

Therefore, a first step will be to link formal methods by techniques used in numerical development of a complex system. Moreover, the tangible advantage of the above methodology is the verification that the analyzed

CHAPTER 2 A STRONG MODELING FOUNDATION

system operates in accordance with the established requirements, and thus a numerical simulation and testing will provide comprehensive coverage of the functions, requirements analysis, and design of the said system.

It is worth noting that, as suggested by the SE philosophy, verification of a generic system presupposes a clear specification of the system of interest, that is, all conditions, related to the execution of the system, must be well understood and well explained so as to form a reliable basis for verification judgments.

A System of Systems (SoS) undergoes continual evolution, as each individual element operates independently and autonomously. This independence leads to a dynamic configuration that constantly changes over time. A systematic engineering process is put in place to evaluate the configuration of a SoS. In each process, the possible spatial and temporal changes of the SoS, known from its previous configuration, are taken into account.

Ultimately, modeling a system is necessary to be able to analyze the evolution of said system at different levels of abstraction.

In order to fully understand the behavior of SoS, there is a need for new and extended theories that can effectively test their automatic, adaptive, and self-awareness properties. These theories should encompass various aspects such as modeling systems as configuration states, considering time-dependent factors, incorporating discrete and continuous elements, and accounting for probabilistic outcomes. This approach allows for a deeper understanding of the complex relationships and dependencies that exist within an SoS, this can facilitate the verification and validation of our model. The unification of modeling techniques would provide a consistent framework for understanding the entire SoS.

For instance, the formal model of the extended aircraft subsystems immediately enables the simulation of different failure scenarios. In this case, the effect of failures on system functionality can be visualized through a graphical interface. This functionality can be used to quickly identify problems in common scenarios before performing a more rigorous analysis.

In the context of a system simulation, measurement plays a crucial role in determining the occurrence of specific behaviors. The measurement process aims to provide a level of confidence for observers regarding the likelihood of a behavior taking place within the system. Each behavior within the system can be associated with a variable, which is represented based on a random distribution. This random distribution reflects the various configurations that the system can assume, depending on its dependencies and interactions. Each measurement carries out an uncertainty that can be evaluated using an analytical approach, for example, the standard deviation is the standard uncertainty for the measurement.

MBSE Advantages

By utilizing MBSE methodologies, system engineers establish a logical sequence or process that clearly defines the tasks to be performed and how they are to be executed. This systematic approach enables engineers to measure various aspects of the system's behavior and performance.

The structure of a process provides different levels of aggregation to allow analysis and definition at various levels of detail to support different decision-making needs. This means that the process can be broken down into smaller components or steps, which can then be analyzed and understood in more detail. At a higher level of aggregation, the process can be viewed as a whole, providing an overview of how different activities are interconnected. This allows decision-makers to understand the overall flow and identify key inputs, outputs, and dependencies. On the other hand, at a lower level of detail, each individual step or component can be analyzed separately. This enables decision-makers to focus on specific aspects or issues within the process.

Therefore, it is necessary to establish a method suitable for performing a task, in other words, it is necessary to define how to perform a particular task.

CHAPTER 2   A STRONG MODELING FOUNDATION

A specific tool is used to accomplish a specific task, applying a specific method. For example, a tool should facilitate the accomplishment of a task, such as making a decision, a tool can be tailored by stakeholders and/or users in a given environment. An environment is, in general, made up of conditions and/or factors that influence the actions of a system. Therefore, the environment must be considered in a way that integrates supporting tools, in other words, the environment then enables (or disables) what to do and how to do it.

A system can be viewed in a virtual working environment, where it is necessary to have a broad or detailed view. Within this environment, we can introduce the methodologies needed to define a task. By simply repeating the same simulation multiple times, the result is likely to be slightly different. The magnitude of these differences gives an indication of the accuracy of the simulation.

When considering the introduction of an XR model for a generic complex system, it's essential to create a virtual environment that closely mimics real-world scenarios. The gap between virtual and real-world environments is the crucial aspect to analyze when designing an XR model that aligns virtual environments with real-world systems. The challenge is creating a complex and engaging system that provides meaningful experiences for its users.

In this scenario the benefits of MBSE are the following:

- Helps to develop and to improve SoS activities
- Improves knowledge of the systems of the SoS
- Clearly traces the relationships between the elements of the subsystems
- Clearly defines the boundaries of the various subsystems

CHAPTER 2   A STRONG MODELING FOUNDATION

- Assesses the impact on the SoS of any changes to the subsystems

- Improves communication between stakeholders in the SoS

- Strengthens the collaboration of teams belonging to different subsystems

- Runs extensive simulations to obtain effective results

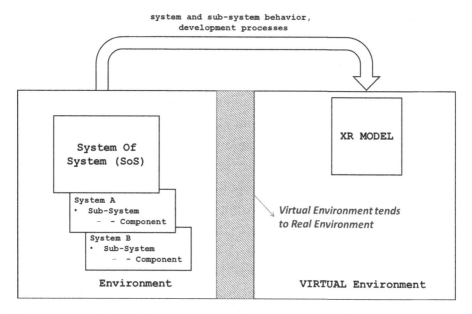

Figure 2-2. *SoS (Real Environment) vs. XR Model (Virtual Environment) in the development phase*

Software Uncertainty

Representations in a virtual environment are affected by errors. Errors are often encountered when the geometry of the represented object does not present its real form in detail or when the virtual environment itself appears to be defined in a limited space. For example an object may appear to be spherical from a certain distance but it turns out to be discretized when it is zoomed in; or suppose we move through a virtual environment and suddenly fall downward aimlessly.

Since almost all augmented reality representations are affected by errors, several methodologies exist in the literature to create transformations by associating the relevant uncertainties. Therefore, we can assume that in almost all XR simulations there will be an uncertainty associated with both the instrument used for the measurement and the repeatability of the performed simulation. In general, uncertainty is part of the system since complete knowledge of its configuration, time-dependent, is often impossible. Suppose to develop a software with certain requirements; uncertainties immediately manifest, if ambiguous or even false requirements have been defined. A requirement is defined as ambiguous when the results of the verification and validation phase can be interpreted in different ways. Therefore, an ambiguous requirement may be considered as incomplete or insufficient, and thus the need to acquire new evidence may decrease the uncertainty. At this point, the environment in which the software will operate must be known, but if non-predictable events come out, the software will not operate adequately for its purpose. Finally, the software must run information, measured, for example, by sensors that may have been calibrated to read inaccurate data or difficult for the software to process. Clearly, when conducting analyses by software, it is not immediate to account for uncertainties, since as the complexity of the system increases, a software will run more and more data obtained from resulting subsystems whose characteristics are unknown.

Ultimately, the lack of knowledge of the above factors will lead to a high amount of uncertainty that may compromise a software's ability to achieve its goals, and it may even cause it to deviate from its behavior. Through the development of technology and the introduction of new models of analysis (i.e., Artificial Intelligence), uncertainty, evaluated during software development, has been managed with the introduction of self-adaptive systems. These are systems that adapt themselves in response to changes, including environmental changes, to elaborate all that information cannot be easily evaluated.

In the development of software systems, it is crucial to define the basic requirements that the system needs to fulfill. However, an often overlooked aspect is evaluating sources of uncertainty that may negatively interfere with these preset requirements. Uncertainty in software development can arise from various factors such as aforementioned incomplete or ambiguous user requirements, changing technological landscapes, or unpredictable external events. It is important to identify and assess these sources of uncertainty early on in the development process to mitigate their potential impact on the system's functionality and performance.

Reproducibility is estimated by running a series of simulations under different conditions, that is, under those conditions that were considered significant for the purpose of the analysis. For example, environmental effects are often significant and sometimes the most difficult to assess because they can lead to nonlinear expressions of the phenomenon being analyzed.

Any software deviation leads to errors that can be included in an uncertainty assessment, so at first it is necessary to quantify sources of uncertainty and then combine them with each other through a statistical approach that provides an uncertainty budget, meaning that each uncertainty is associated to a sensitivity value. The standard uncertainty for each source can then be calculated and finally combined.

Each source of uncertainty involves the same uncertainty of the measurement result. Introducing the sensitivity coefficients, let us begin by considering the simple case in more detail. It is important to remember that it is impossible to know the true value of an object we are measuring – we can only know the result of a measurement. This measurement will have some error, which is the difference between the true value and the result of the measurement. The uncertainty of the result of a measurement and the uncertainty of the error are mathematically equivalent. We will consider these unknowable quantities, the true value and the error, in a theoretical analysis. Let us now define the simple case in which each source of uncertainty involves an equivalent uncertainty in the measurement result.

Monte Carlo Simulation

When no measurements are available, it is possible to generate input data to the system from the original data through some statistical variation so that the result can be compared with these original values.

To overcome problems of measurement uncertainty and accuracy, Monte Carlo simulations are often used. These simulations have the advantage of operating on multiple input vectors (high dimensionality) and provide statistically significant results to estimate the quality and accuracy of the measurement.

In this chapter, we focus the attention on the Monte Carlo simulation method often used in MBSE; they can be defined as a set of interrelated processes and tools. Thanks to the evolution of technology and the development of increasingly complex software, simulation methods have found massive applications in recent years to verify and validate models.

Simulation (from the Latin similis) refers to the creation of an abstract model capable of replicating the behavior of a system, phenomenon, or individual process. In general, these are procedures by which mathematical models are constructed, on which probabilistic calculations can be performed, based on sampling, or differential calculations, based on reasonable approximations.

In the use of virtual reality (VR), it is clearly important to start with the development of a three-dimensional (3D) model. Where we do not have tools available to measure digitally the actual subsystem, we can base our analysis on methods recognized as valid in the MBSE literature. The Monte Carlo method, a powerful statistical technique, holds

CHAPTER 2   A STRONG MODELING FOUNDATION

tremendous potential for enhancing the metaverse experience. By utilizing this method, accurate simulations can be reproduced, providing users with more realistic and engaging virtual environments. This technique allows us to replicate a random reality, without relying solely on digital measurements. By employing this method, we can simulate various scenarios and gather insights that aid in understanding and improving complex systems within virtual reality and the metaverse.

While digital measurement tools may have limitations in certain areas, leveraging methodologies recognized as valid in the MBSE literature offer a robust alternative for analysis and decision-making. The Monte Carlo method serves as an effective tool in reproducing random realities within virtual environments, facilitating enhanced training experiences and advancing our understanding of these dynamic systems.

We will see better with the example shown that follows (aircraft maintenance) where the environment (aircraft), surrounding the model to be represented (failures on hydraulic system, autopilot, or fuselage), can be based on VR technology. In this system, the scene can be designed in accordance with reality, which represents the reunification of VR technology and engineering applications. It is shown that by using VR technology in simulating the consequences of aircraft failure, the whole system has visual, realistic, and real-time characteristics, which not only embodies the great engineering value of virtual reality technology but also provides references for predicting the impact of failure.

Monte Carlo methods rely on sequences of pseudo-random numbers or make use of any procedure that produces random numbers. Monte Carlo simulation is an extremely useful and versatile technique for understanding variations in production processes and measurement of uncertainty. The Monte Carlo method is useful for analyzing the propagation of uncertainty, and determining how lack of knowledge affects the sensitivity, performance, or reliability of the aforementioned "maintenance system."

CHAPTER 2 A STRONG MODELING FOUNDATION

The Monte Carlo method dates back to the studies carried out by Fermi and Ulam, who wanted to calculate the area of a function f(X) between the extremes a and b belonging to X, that is, to calculate the integral of the function defined between a and b. In order to avoid the analytical approach, that is, to calculate the said integral by canonical methods, the idea of the two scientists was to find the maximum value y_{max} of the function f(X) in the considered interval, and then construct a rectangle of height y_{max} and base $|a - b|$. This approach was consequently refined by Von Neumann, who indicated the usage of statistical inference to calculate the value of the integral to a certain approximation. The computation is initialized with a counter that randomly generates a huge number N of pairs of numbers (x,y), and at each iteration a logical 0, 1 instruction must occur to figure out whether the pair has fallen inside the area of interest (e.g., 1) or remains outside (e.g., 0). The number of points that fall within the area (e.g., a quarter of a circumference shown in Figure 2-3) of interest will be N_i.

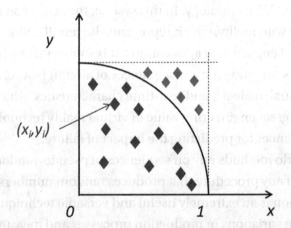

Figure 2-3. *"Heat" or "Miss." How many points heat the target?*

Using a trivial proportion, the area searched will be:

$$M \approx \frac{y_{max} |a-b| \cdot N_i}{N}$$

CHAPTER 2 A STRONG MODELING FOUNDATION

If we know exactly the value $y_{max}|a-b|$, then we define the probability that a solution M is contained in y_max |a-b|, equally, if we can count all pairs (x,y) for a total of N attempts, then we define the probability as:

$$p = \frac{N_i}{N} = \frac{M}{A}$$

with $A = y_{max}|a-b|$.

Monte Carlo simulation is a statistical analysis method that can be very useful in all those circumstances where decision-making is based on reliable and accurate data, such as in assessing the risks associated with a project, such as evaluating the time and cost of different activities. Let us begin by defining the general scheme of Monte Carlo simulation; it is generally divided into the following functions:

- Generate random values that define a measure of the predefined activity (i.e., maintenance cost, number of failures, etc.).
- Count the total Ni amount of random measurements that fall within the area of interest.
- Calculate the probability of the measurement known the total number N of iterations performed.

Therefore, generating a sample of one or more random variables is an essential ingredient of any Monte Carlo experiment. The distribution of random variables can be treated by different approaches such as (and not limited to) the following:

- Random variable distributed according to Bernoulli
- Random variable distributed according to Weibull

CHAPTER 2   A STRONG MODELING FOUNDATION

- Uniformly distributed random variable in a given interval
- Exponentially distributed random variable
- Random variable with constant density at intervals

On the y-axis, the function f(X), with X a random variable, called the cumulative distribution function, belongs to the interval [0,1]. Therefore, to derive X, it is sufficient to calculate the inverse of the function f(X). In addition, the areas subtended by a given interval on X allow us to know the probability of occurrence recorded in the interval of X considered.

One of the key applications of the Monte Carlo method in the context of the metaverse is within Model-Based Systems Engineering (MBSE). MBSE involves creating digital models that simulate real-world systems, enabling engineers to design and optimize complex systems efficiently. By employing Monte Carlo simulations within MBSE for metaverse purposes, developers can ensure that virtual worlds accurately reflect real-world scenarios.

The use of Monte Carlo simulations enables designers to incorporate factors such as physics-based interactions, environmental effects, and realistic behavior patterns into these virtual experiences. Leveraging the power of Monte Carlo simulations for metaverse purposes holds immense promise. By striving for accuracy and faithfully reproducing real-world scenarios within virtual reality environments, we can create an even more immersive and captivating metaverse experience for users worldwide.

In conclusion, it is clear that the Monte Carlo method must involve a large number of simulations that can be summarized into a final result that is a function of the variables distribution. A virtual reality product that can integrate several simulation methodologies shows that performance with repeated use is better and easily measured.

MBSE Applications

There are many methodologies for evaluating the performance of a complex system. Particularly since we talk about dynamic data, which vary over time, that shall be optimized by particular functions so that the system achieves its intended goal. Often, some analysis is carried out to determine the cost and execution time of all those processes that are interactive and affect two or more functions simultaneously. The ability to discern the most relevant information is an essential component of the ability to anticipate variation in all those variables considered in a model. Because of the technological limitations of current software, a traditional method for exploring the behavior of complex systems cannot be reliable. Through technological improvements, virtual reality (VR) can overcome these limitations by providing a massive amount of numerical simulations within immersive, interactive environments.

MBSE is necessary to effectively design and develop complex systems, which from this point onward shall be considered as analyzed systems within virtual frameworks (see Chapter 4). The methodologies used to prepare a model and realistic simulations may still be difficult to understand. Certainly, the MBSE approach and its techniques can facilitate the verification of the virtual reality model. This verification effort is normally carried out by subject matter experts (SMEs), who in turn verify the requirements against the stakeholder needs.

From an industrial metaverse perspective, the degree of detail of a model needs to reflect the customer's purpose for the system (e.g., demonstration, co-design, analysis, etc.). So it is essential to define its parameters very precisely, which come from other systems that are previously simulated and analyzed. Therefore, each analyzed system can be modeled from a set of random input variables, which are characterized by a probability distribution. The output variables, dependent on the random inputs, will also be random variables. However, since the probability distribution of the outputs is not known, the objective of simulating the system is in fact to determine the behavior of the outputs. The model can be supplemented by a mathematical model that determines how the values of the outputs are calculated based on the values of the input variables.

The model effectively becomes the main artifact of the MBSE and can be represented first as graphical diagrams (e.g., SysML) and later through 3D visualizations in a virtual environment. These allow all stakeholders to tap into relevant information about the system and will ultimately enable decision analysis, economic analysis, and risk analysis.

In the field of maintenance, experience plays a crucial role in determining the appropriate actions to be taken. However, relying solely on past experiences may not always result in optimal outcomes. To carry out simulated maintenance, we rely on MBSE to create a 3D framework that includes all the necessary information to render the system as well as its user interface. A VR application shall provide the user with a real-time simulation and engage multiple sensorial channels.

MBSE tools include concurrently multiple system modeling methods (e.g., SysML models, probabilistic models as well as agent-based simulation or model-based storytelling). In addition, it is always appropriate to model not only the behavior but also the processes of the system life cycle. The virtual system model can range from light models to full models. Lightweight models reflect simplified structure (e.g., simplified geometry) and simplified physics (e.g., reduced order models) to reduce computational load, especially in early design activities. These lightweight models allow complex systems and systems of systems (SoS) to be simulated with fidelity in the appropriate dimensions to answer questions with minimal computation.

From the physical twin, performance, maintenance, and health data can be collected and provided to the digital twin. These data include operating environment characteristics, engine and battery status, and other similar factors. The digital twin and physical twin can be supported by a shared MBSE repository, which also supports SE and data collection tools. MBSE models constitute the authoritative sources of truth. This configuration also ensures bidirectional communication between the digital twin and the physical twin.

The MBSE approach applied to the philosophy of virtual reality tells us that any system, while not corresponding to reality, can be managed and considered virtually for the purpose as potential developments in representation or logical inference.

In the example, shown in the next section, a VR model will be analyzed using a functional diagram. The distribution of input will lead to defining a relevant distribution of output as time and cost of each maintenance action, which in turn could have a normal or uniform distribution. This model can be developed as a textual/augmented VR (interaction, no immersion) and/or a desktop VR (interaction, immersion) and/or an immersive VR (interaction, high immersion). To a first approximation, a normal pattern may be assumed without compromising the result. In the example, it will be assumed for simplicity that the cost of each action is independent of that of the others, the total cost of the project also constitutes a variable whose value lies in a range between a minimum and a maximum. This variable will be normally distributed as a sum of random variables. For this reason, it is not very important to define precisely the type of distribution of the cost of each action. In some cases, the cost of a maintenance action is certain and is fixed. This is also a simplification, as in reality, an activity has an estimated cost that varies within a specific range. Similar reasoning is used to estimate the time required to perform each activity.

A Small Practical Example – Overview

Have you ever found yourself receiving a lot of information and not knowing how to put it together? Well, the more we are immersed in a multidisciplinary activity or operate in a complex system, the more we need to bring order to it. In our heads, we build functional or procedural

CHAPTER 2   A STRONG MODELING FOUNDATION

models that actually depend on both our knowledge and our ability to process so many dependent variables. Now, imagine that you are in a virtual environment (VE) that should contain all the information needed to perform the task in the best possible way. Such an environment must be constructed in a manner consistent with the real environment that you want to "simulate." So far, no problem with having tools that generate 3D models or collect a lot of data. Unfortunately, the environment is not static, but is subject to constant change. This is because such an environment should allow the user to interact with it as if it were interacting with the real thing.

Creating a logical flow chart for maintenance activities can be a valuable tool in setting up a Monte Carlo simulation. By detailing the steps in maintenance processes, the variables and their distributions needed for simulation can be correctly defined. By clearly defining each step and the associated probabilities, you can better model the uncertainties in maintenance processes and predict their impacts on equipment performance and reliability.

If we take as example the maintenance system of airplanes (also called aircraft in engineering jargon), as shown in Figure 2-4. The maintenance is performed on an aircraft, which has accumulated $N=12,500$ flights, landed in North America at an average temperature of 20 °F.

In general, a maintenance process supports the in-service system in order to sustain its capability, validated and verified in the V&V stage. For simplicity, we consider three systems belonging to the SoS: aircraft, maintenance, and logistics.

CHAPTER 2   A STRONG MODELING FOUNDATION

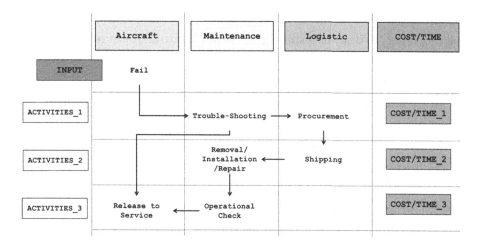

Figure 2-4. *Logical Flow Chart of maintenance activities*

With the use of XR, we can figure out two scenarios for aircraft maintenance. In the first, one or more failure cases can be simulated and the user can try his hand and practice (virtual) maintenance.

In such a scenario, main maintenance activities, as mentioned in the INCOSE SE handbook, can be performed in order to restore system operation after an occurred failure. In particular, maintenance procedures for preventive and corrective maintenance can be developed and/or future maintenance by reviewing and analyzing anomaly reports can be planned.

In the second, the user can interface directly with a real system (the aircraft) remotely and perform maintenance procedures defined in the design stage. In this case, presuming that the aircraft can transmit enough relevant real-time information about its status, the user will be able to perform remote troubleshooting (e.g., telepresence) and implement the necessary maintenance actions. In this scenario, the user can analyze and correct any maintenance action to remedy previously undetected design errors.

45

CHAPTER 2   A STRONG MODELING FOUNDATION

The aircraft could inform the user of the overt failure on the upper assembly (e.g., hydraulic system) and then in a virtual room represent both the 3D model of the mechanical failure and the wiring diagram (also a model) to simulate the eventual electronic failure.

The MBSE methodology enables a clear definition of an "augmented maintenance system" in order to avoid waste and overcome the limitations of failure assessment based on traditional 2D models.

For the first scenario, that is, using XR to simulate aircraft maintenance, it might be possible to create a library of reusable faults that can be applied to different mechanical or digital components of the system. Despite limitations due to system complexity, this process would be integrated into existing on-board tools, allowing the system engineer to choose from a palette of predefined failures and drag and drop them onto existing system components as they go. A similar logic is used in flight simulators.

The physical maintenance activity could be performed on site or through an XR with support of cobots, that is, drones equipped with sensors, which in real-time acquire data and transfer them to the model. The cobots could provide a useful dataset that takes account of the number of completed use cases, the number of requirements met, and the percentage of logical components.

With appropriate modeling (simulation and diagrams), even the logistics of aircraft maintenance can be managed remotely.

At this point, it is essential to prepare a coherent architecture of the SoS that will take into account different modeling languages capable of expressing multiple, independent, and interconnected interactions. In addition, the modeling language chosen must take into account the possible inoperability of one or more constituent parts of the SoS. Ultimately, the connections and interfaces between the systems considered must take into account the main aspects of each, so as to generate a similarity consistent with the real case.

This model can be subjected to various types of analysis, such as completeness and consistency analysis, model verification, and theorem testing. Having a formal model of the system, extended to the failure model, different failure scenarios can be simulated. This is an important feature, since engineers can visualize the effect of faults on system functionality while monitoring their activation through a graphical interface. As we mentioned before, this functionality can be used in common scenarios before performing a more rigorous static analysis.

We have decomposed the process into three elementary subprocesses, which can be easily represented in numerical or mathematical terms. Three possible CASES, associated with three distinct failures, are listed below (consider even more than three with relative increase in model complexity) to assess the cost and timing of the activities to be performed to release the aircraft into service after maintenance:

- CASE 1: Autopilot failed
- CASE 2: Hydraulic fail
- CASE 3: Structural Damage (Bird Strike or Lightning Strike)

The norm ARP4761 1196-12 – GUIDELINES AND METHODS FOR CONDUCTING THE SAFETY ASSESSMENT PROCESS ON CIVIL AIRBORNE SYSTEMS AND EQUIPMENT, describes the differences between failure, fault, and error.

- Failure: A loss of function or a malfunction of a system or a part. A failure is an event that occurs when the behavior of the system in service deviates from the correct way of operating. Where in-service behavior is associated with a series of actions/states, observed from the external side, then a failure means that at least one or more actions/states of the system deviate from the correct service state/action.

CHAPTER 2　A STRONG MODELING FOUNDATION

- Fault: An undesired anomaly in an item or system.

- Error: An occurrence arising as a result of an incorrect action or decision by personnel operating or maintaining a system. The error is the effect generated by the fault. In most cases, a fault first causes an error in the service state of a component that is part of the internal state of the system, and the external state is not immediately affected. Note that many errors do not reach the external state of the system and do not cause a failure.

In general, the failure model can be formed down to the lowest level of the component hierarchy, as a single component failure can propagate and affect the operation of other connected components. Lately, engineers have been relying on all those processes necessary to prevent failures during the expected life of the system. These approaches involve predicting failures and avoiding failure modes by specifically improving the system reliability. Through MBSE's methods, it is possible to load into the virtual room all the documentation used in the design phase and validated during aircraft certification. For example, it is possible to load into virtual reality the system architecture and FMEA (Failure Mode and Effect Analysis) documentation that was used to generate the fault tree. A fault tree generated by a traditional manual analysis is usually more intuitive to read because the analyst creates the fault tree that corresponds to the structure of the system.

The fault trees generated by any tool depend entirely on how correctly the functional requirements have been defined and especially on detailed analyses of possible loss of functionality for both the system and its model. If all possible failures have not been considered in the development phase or if component specifications do not match the behavior of the in-service system, fault trees may be incomplete or inaccurate.

Therefore, in this particular case, engineers must determine the causes of occurred failures and apply new methods to estimate the likely reliability of the in-service system.

A Small Practical Example – Simulation

1. The failure on the autopilot is imputable to an automatic disconnect that depends on loss of control and lack of maneuvering

    - On the pitch or roll by elevators or ailerons
    - On the automatic throttle, which allows the aircraft's thrust to be controlled automatically
    - On the yaw by the rudder controls only during approach

2. Failure on the hydraulic system can be traced to the following root causes:

    - Hydraulic pumps process fluid outside allowable values.
    - Empty hydraulic fluid reservoir.
    - Hydraulic system pressure is lower than allowable.

3. Damage caused by a bird strike or lightning strike can be assessed following an inspection of the aircraft structure: fuselage, wing, and tailplane. Structural damage that could be identified are generally:

    - Dent
    - Hole
    - Delamination (separation of layers in the material)

CHAPTER 2   A STRONG MODELING FOUNDATION

Given the mean and standard deviation, we can propose a simulation describing the probability of occurrence of a particular failure so that the right measures can be taken in an evidential environment at 20 °F. Once the possible failure is identified, following trouble-shooting, the necessary maintenance intervention can be simulated. If, on the other hand, we want to perform the trouble-shooting simulation according to a random variable, we apply the Monte Carlo method, considering the special case of the hydraulic system failure.

Given the previous failures, occurring at a certain number of flights and recorded in appropriate dedicated reliability databases, we derive the mean and standard deviation as shown in the Table 2-1.

Table 2-1. *Statistical parameters obtained from failures database*

ITEM	AVERAGE (μ)	ST.DEV (σ)
Pump	1537.5	1120.8
Hose	4033.3	895.4
Reservoir	1049.4	171.4

We choose to simulate the event, running more than 1000 trials, by introducing a random variable belonging to the interval [0,1]. For each random value of the variable

1. The potential failing item is associated (Table 2-2).

CHAPTER 2   A STRONG MODELING FOUNDATION

Table 2-2. *At each trial, a failed item is associated to variable random*

TRIAL	Variable_Rnd	Fail_cutoff
1	0.13750	Pump
2	0.81705	Reservoir
3	0.47353	Hose
4	0.95517	Reservoir
5	0.98268	Reservoir
6	0.81162	Reservoir
7	0.03089	Pump
8	0.24873	Pump
9	0.60125	Reservoir
10	0.15754	Pump
...	...	...
1168	0.85671	Reservoir
1169	0.76285	Reservoir

2. The number of cycles at which the failure occurs is calculated by applying the inverse of the following Gauss function (Table 2-3), with x cycles, μ mean, and σ standard deviation:

$$f(x) = \frac{1}{\sigma\sqrt{2\pi}} e^{-\frac{1}{2}\left(\frac{x-\mu}{\sigma}\right)^2}$$

CHAPTER 2   A STRONG MODELING FOUNDATION

Table 2-3. *Cycles estimation applying inverse of Gauss' function*

TRIAL	Inv _pump	Inv _hose	Inv_reservoir
1	314.04	3055.92	862.26
2	2550.93	4842.96	1204.35
3	1463.09	3973.89	1037.98
4	3439.70	5552.99	1340.27
5	3905.36	5925.01	1411.48
6	2528.15	4824.75	1200.86
7	-555.90	2360.93	729.22
8	777.06	3425.82	933.07
9	1825.09	4263.09	1093.34
10	411.55	3133.82	877.17
...	...	...	...
1168	2731.89	4987.52	1232.02
1169	2339.43	4673.99	1172.00

According to the proposed simulation, it turns out that failure on the reservoir is more likely (Table 2-4). The simulated probability, mean, and standard deviation are calculated by the following formulas, respectively:

$$p = \frac{N_i}{N}, N_i = count\ of\ single\ item, N = 1169$$

52

Table 2-4. Simulation results

ITEM	SIM_PROBABILITY (p)	SIM_AVERAGE (μ_sim)	SIM_ST.DEV (σ_sim)
Pump	30.1%	1527.3	1134.9
Hose	29.2%	4025.1	906.7
Reservoir	40.7%	1047.8	173.6

Table 2-5. Error estimation after comparison between average and standard deviation

ITEM	Err_AVERAGE	Err_ST.DEV
Pump	0.67%	-1.26%
Hose	0.20%	-1.26%
Reservoir	0.15%	-1.26%

Therefore, a SCENARIO can be associated with each activity (training or practical) in which the model allows the various subsystems to exchange information with the outside world once they have interacted with each other, and vice versa. This model makes it possible to input data and obtain results without having to vary its structure. It is clear that we can easily switch from one subsystem to another by entering information that is useful for the final result. For example, in our case, to estimate times, we need to draw on the database that contains the execution times of each individual activity related to the type of failure (aggregation).

Therefore, it is imperative for analysts to diligently gather information from various sources and employ appropriate methods that align with the specific requirements of the analysis. This approach ensures a more robust and reliable analysis, reducing the likelihood of failure and enhancing overall decision-making capabilities.

CHAPTER 2   A STRONG MODELING FOUNDATION

We could have two possible scenarios to study the behavior of the system under the established failure conditions.

- SCENARIO 1: On-site maintenance and activity management (including logistics) with a virtual model
- SCENARIO 2: Remote maintenance with a virtual model (training no logistics)

The proposed simulation is particularly effective because the evolution of the system, consisting of an extremely large number of elements, can be reduced to the combined effect of simple interactions among the various elements. As we have seen, one approach to achieving accurate simulations is through the use of random simulation techniques, such as Monte Carlo analysis. By incorporating randomness into our models, we can capture a wide range of potential scenarios and outcomes, enhancing the realism and robustness of our simulations; therefore, we can create XR simulations that are fit for purpose in industrial settings.

Finally, it is important to underline that the usage of "models for simulation for system behavior," while not accurately representing real-world scenarios, is critical for industrial applications, without forgetting that the level of realism may vary depending on the specific application and goals.

Modeling Languages

Software engineering has given us literally hundreds of modeling notations. Modeling is used extensively in engineering, solves issues connected to design, and will typically use simulation. MBSE combined with XR technology offers a powerful approach to facilitate the development of product life cycle-specific XR models. This approach enables stakeholders such as product designers, production managers, customers, and investors to gain a better understanding of the product's behavior in later phases. By leveraging MBSE techniques, which involve using a formalized language for system modeling, and integrating XR technology into the process, stakeholders can visualize and interact with virtual representations of the product.

Modeling languages must simplify the behavior of a real system and represent the capability, functionality, and performances of the real system. To effectively represent changes in these systems, it is important for modeling languages to be flexible. This flexibility allows for dynamic reconfiguration of the model, accommodating modifications as needed. By supporting dynamic reconfiguration, modeling languages enable constituent systems to upgrade their interrelationships. This means that as the system evolves or requirements change, individual components can adapt by establishing new connections or terminating existing ones. Overall, the ability of modeling languages to facilitate dynamic reconfiguration ensures that models accurately reflect the evolving nature of complex systems. In addition, the modeling languages need to be robust enough to allow for efficient optimization and fault management.

Resuming the example shown above, it is important to emphasize in the context of modeling languages that a key task of the system engineer is to receive all information about the dependence of the fault order and consequently allow the analyst to create a synchronization (requirement for XR). In order to accurately define all fault propagations, XR needs to simulate all possible cases, including simultaneous/dependent and persistent/intermittent scenarios. However, it's important to note that controlling the triggering of faults may not always be possible. Therefore, one must plan to introduce the fault either manually or through a server input in real-life situations. This could obviate the non-deterministic fault generation, in fact the simulation, as we have already seen in the previous paragraphs, is conducted on a statistical or at most random basis. This ensures that all potential faults are accounted for and properly tested in XR simulations.

CHAPTER 2   A STRONG MODELING FOUNDATION

In addition to what is said above, emergent behaviors are fascinating phenomena that occur when simpler elements or systems interact, resulting in complex patterns or behaviors. In the realm of XR, these emergent behaviors play a crucial role in creating immersive and multisensorial experiences. The complexity of emergent behaviors in XR is a result of the dynamic interactions between users, virtual objects, and the environment. As users navigate and engage with XR experiences, their actions can trigger unexpected responses or outcomes. These emergent behaviors can range from subtle changes in object behavior to entirely new interactions that were not explicitly programmed.

Understanding and harnessing emergent behaviors in XR opens up exciting possibilities for creating more immersive and interactive experiences. Developers can design systems that allow for user agency and exploration while still maintaining a sense of coherence within the virtual environment. However, it is important to note that managing emergent behaviors in XR poses challenges as well. Ensuring that these behaviors enhance rather than detract from the user experience requires careful design considerations and testing.

Takeouts

- The introduction discusses the use of Virtual Reality (VR) technology for industrial purposes, focusing on the representation of the real world through models. Virtual objects and environments are the foundation for creating immersive experiences, and digital representation is essential. The construction of models is determined by experience and verification, and the MBSE approach can be integrated into the XR technology design process. MBSE aims to improve system engineering by using models to help designers

CHAPTER 2   A STRONG MODELING FOUNDATION

better understand and visualize complex systems. It is used in various industries, including defense, aerospace, and software development. A single system model serves as a "source of truth" for all systems analysis tools, allowing for faster updates and better understanding of system performance. Understanding and analyzing systems within XR allows for more effective and impactful applications. Creating a system model involves understanding its engineering and design, simulating real-world scenarios, and testing performance before implementation.

- System engineering involves programming systematic activities for simulation and reproduction of a model to optimize projects. Model-Based Systems Engineering (MBSE) is an emerging approach in the field of SE that formalizes the application of modeling principles, methods, languages, and tools to the entire life cycle of complex, interdisciplinary socio-technical systems. MBSE focuses on the boundaries and interactions between seemingly independent and evolving systems. To create a model-based system, informal design models and requirements documents must be considered. More accurate models must be sought from the system architecture, which defines the behavior of the system and automates parts of the analysis process. Proper system architecture allows for simplified and customized model views, which can be validated through simulation. However, the additional complexity due to multiple system life cycles and balancing different requirements affects the performance of the various systems constituting the SoS.

CHAPTER 2 A STRONG MODELING FOUNDATION

- Virtual reality (VR) can be used for engineering purposes to imitate and reproduce the behavior of a particular observed system. However, individual consciousness can influence virtual representation, leading to different levels of abstraction. To prevent biased perceptions, a correct simulation is necessary to ensure accurate choices during engineering activities. Numerical simulations are useful for analyzing complex systems, as they provide artificial but concrete models. Verification of generic systems requires a clear specification of the system of interest. System simulations help analyze the evolution of systems at different levels of abstraction. New theories that test automatic, adaptive, and self-awareness properties are needed to fully understand system behavior. Measurement plays a crucial role in determining the occurrence of specific behaviors in system simulations. MBSE methodologies provide a logical sequence for defining tasks and execution, allowing for analysis and definition at different levels of detail. The virtual working environment allows for the introduction of methodologies needed to define tasks. Benefits of MBSE using VR include improving knowledge of systems, tracing relationships, defining boundaries, assessing impacts, improving communication, strengthening collaboration, and running extensive simulations.

- Software uncertainty is a significant issue in virtual environments, as it can arise from various factors such as incomplete or ambiguous user requirements, changing technological landscapes, or unpredictable external events. To mitigate this uncertainty, it is crucial to identify and assess these sources early on in the

CHAPTER 2   A STRONG MODELING FOUNDATION

development process. Monte Carlo simulations are often used to overcome measurement uncertainty and accuracy issues. These simulations operate on multiple input vectors and provide statistically significant results to estimate the quality and accuracy of the measurement.

- In the use of virtual reality (VR), it is essential to develop a three-dimensional (3D) model to generate input data for the system. By utilizing the Monte Carlo method, accurate simulations can be reproduced, providing users with more realistic and engaging virtual environments. This technique allows for the replication of random realities without relying solely on digital measurements, facilitating enhanced training experiences and advancing our understanding of dynamic systems within virtual reality and the metaverse.

- The Monte Carlo method is a statistical analysis technique used to understand variations in production processes and measurement of uncertainty. It is useful for analyzing the propagation of uncertainty and determining how lack of knowledge affects the sensitivity, performance, or reliability of a maintenance system. The method is based on sequences of pseudo-random numbers and can be applied to various situations, such as assessing risks associated with a project or evaluating the time and cost of different activities. In the context of the metaverse, Monte Carlo simulations are crucial for Model-Based Systems Engineering (MBSE), enabling engineers to design and optimize complex systems efficiently. By incorporating factors such as physics-based interactions,

59

environmental effects, and realistic behavior patterns, virtual reality environments can create immersive and captivating experiences for users worldwide.

- MBSE applications are crucial for evaluating the performance of complex systems, especially those with dynamic data. Traditional methods are insufficient due to technological limitations, but virtual reality (VR) can overcome these limitations by providing numerical simulations in immersive environments. MBSE helps design and develop complex systems, which can be analyzed within virtual frameworks. The model becomes the main artifact of MBSE, representing graphical diagrams and 3D visualizations in a virtual environment. In maintenance, MBSE creates a 3D framework that includes all necessary information and engages multiple sensorial channels. MBSE tools include multiple system modeling methods and can range from lightweight models to full models. The MBSE approach applies to the philosophy of virtual reality, allowing any system to be managed and considered virtually for potential developments in representation or logical inference.

Questions

- Consider the case of autopilot failure and according to a Gaussian probability, with an average equal to 2400 cycles and a standard deviation equal to 400 cycles, calculate the total time of activities for putting in service the aircraft.

CHAPTER 3

XR Life Cycle

The best way to predict the future is to create it.

—Peter Drucker

In this chapter, the XR life cycle is described as an iterative framework, integrating verification and validation processes throughout to ensure high-quality outcomes that meet user expectations and industry standards. This chapter focuses on the holistic approach that enables developers to create immersive and effective XR experiences. Therefore, identifying the purpose and objectives of the XR application, the Extended Reality (XR) life cycle is described as a process that involves the development, deployment, and maintenance of XR technologies, including Virtual Reality (VR), Augmented Reality (AR), and Mixed Reality (MR).

The goal is to provide a comprehensive explanation of how a generic product can seamlessly interact within a dedicated environment. This interaction goes beyond simply representing design scenarios; it encompasses the entire iterative process of product development, while considering all aspects of the life cycle. Drawing upon the guidelines provided by the INCOSE handbook, we will delve into the various stages and considerations involved in product development. From conceptualization to implementation, and from production to disposal, every step in the life cycle will be carefully examined.

CHAPTER 3   XR LIFE CYCLE

The Life Cycle of Technology

Since the late 1980s, all major design activities (Specification Analysis, Feasibility Study, Logic-Functional Design, Testing, etc.) of a generic product have been carried out concurrently in order to produce a product with the highest level of quality and performance, while minimizing costs, so that the shortest time-to-market can be achieved. This approach currently has reason to exist because it is aided by increasingly rapidly evolving information technology. The integration of computer tools into design activities, for example, Computer Aided Design (CAD) support and computer simulations, has inevitably led to qualifying personnel to use these technologies. Moreover, the aforementioned tools are predominantly used in the product development phase. Therefore, VR aims to solve the possible limitations of current computer tools, starting from the simple visualization of 3D models to the dynamic and interactive representation of the product.

To limit our discussion to XR product development is to limit the benefit of this technology to a single phase without considering, for example, the product service and disposal phase. If we think about it for a moment, making the entire system life cycle in a VR environment can help us meet industrial requirements related to cost, quality, and time (Figure 3-1). The term "virtual" (final form of a design concept is not yet generated), when applied to the representation of an entire industrial system (dedicated to the realization of a product), may lead to focusing solely on digital mock-ups, hence neglecting the requirements of each product development phase. A VR representation of different phases of a system's life can help the system engineer get an "outside" look, thus helping to recognize and understand missing requirements. As shown in Figure 3-1, starting from the concept of a complex system, we can associate the VR model with all systems defined in the literature (PHYSIC, CONCEPTUAL, NATURAL, and HYBRID). Ultimately, XR serves as an invaluable tool for visualizing concepts, simulating scenarios, testing functionality, and identifying potential issues throughout the entire life cycle of a product. Its ability to immerse designers and users into virtual environments fosters better comprehension, making it an indispensable asset in modern-day product development practices.

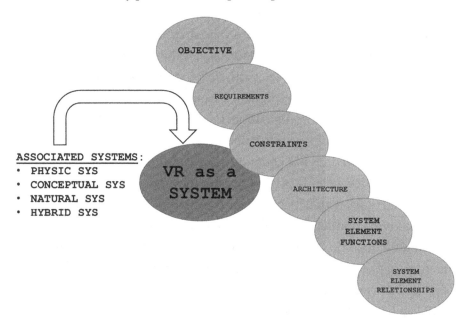

Figure 3-1. *To associate all existing systems to VR/XR*

System Life Cycle and XR

When considering the life cycle of a system, it is interesting and particularly appealing for a system engineer to introduce XR in support of the main stages of the life cycle. In this chapter, we will show how XR can easily facilitate the activities contemplated in each stage to integrate the methodologies presented in SE literature. To get an overview of what will be discussed, let us consider, for example, the maintenance activities contemplated during the service phase of a generic system. During the design and production phase of a product, careful consideration is given to the maintenance phase as an essential aspect of its life cycle. The goal is to anticipate any potential issues that may arise and assess the associated losses in terms of time, resources, and costs. Maintenance costs, in particular, constitute a significant percentage of the overall life cycle expenses.

What should be done to propose a valid maintenance model during the design and production phase? Trivially, a valid maintainability model is mainly based on full-scale physical prototypes of the product. Therefore, according to this approach, it would be advisable to always create a prototype of the product. What happens if a physical prototype is difficult to make? Again, are we sure that the establishment of a maintenance model does not generate negative effects on design? To date, maintainability design work is usually done through computer-aided design (CAD) tools, with which designers can produce maintenance animations and conduct simulation-based design analyses. However, this work is time-consuming, laborious, and requires experienced personnel. Therefore, one wonders what are the benefits of virtual modeling for design and consequently for maintainability?

XR can be seen as a logical progression from traditional CAD tools, by using XR technology, designers are able to interact with these objects within a virtual environment (VE). This interaction takes place through various interaction devices that enable designers to manipulate and explore their designs in a more immersive and intuitive way. Designers can now have a more tangible experience with their creations in the VE, enabling them to make well-informed decisions regarding maintenance activities for the final product.

How can we introduce into a VR environment all the information needed to define the life cycle of a system? Let us revisit for a moment what was discussed in Chapter 2 to see if it is possible to devise a model-based approach. In fact, MBSE is a systematic and holistic development approach that uses seamless system modeling, starting from stakeholder and requirements specifications to system integration, verification, and validation. The advantages of using MBSE certainly include vertical and horizontal traceability, but most importantly, the reuse of models in different environments and scenarios.

How can all the modeling requirements be brought together? A common modeling language is SysML (Systems Modeling Language) because this language allows the model repository to be partitioned from geometric views for possible analysis. The repository contains the overall model with all elements and connections. The views, on the other hand, show only the context-specific elements to reduce the complexity of modeling and analysis. Views in SysML are realized in terms of diagrams divided into structural, procedural, and requirements diagrams.

Integrating various modeling languages has become a priority for many enterprises as they strive to establish connections between model representations through functional diagrams and 3D models.

By leveraging VR-enabled SysML (Systems Modeling Language), enterprises can bridge the gap between physical prototypes and their digital counterparts, leading to improved collaboration, visualization, and decision-making throughout the product development life cycle.

XR Applications in a System Life Cycle

As we mentioned in the introduction to this chapter, models in XR can be manipulated by the user in a given environment through appropriate interactions. Again, the different phases of the life cycle can be built within an XR environment by implementing all the procedures, techniques, and methodologies related to the phase under consideration. All of these elements must be provided not only with geometric, but more importantly, procedural descriptions. While current research and development in XR mainly focus on improving visual presentations and increasing the quality of content, it is necessary, when talking about the system, to take into account all the aspects of the system life cycle. Therefore, XR can represent all activities pertaining to different stages of the life cycle.

Currently, there is no acknowledged method for building procedural models in XR. All that can be done is based on manually adding simulations. Therefore, there is a need for a new efficient method for the proper configuration of the XR scenario in which to insert the procedures pertaining to the objectives of the stage being examined.

Thus, the virtual environment must render

- The correct representation of the scene in which to work (even for complex systems)
- Reusable models (in different scenes), also called assets in XR jargon
- Extended user interaction

In addition, as discussed earlier, an XR scene should include models of environments in which all processes related to the life cycle phase can be "virtually" applied. This places additional requirements on the new model, which must

- Be able to describe parts separately from each other, with appropriate interfaces to link them together
- Allow for interactions between the parts and interactions with the user, so as to achieve an interactive scene

Let us try at this point to create a standard process for creating a life cycle model of a system in XR. Considering the above requirements, we can meanwhile choose SysML as the language for procedural descriptions in XR. Any model that possesses geometry is naturally introduced according to the 3D logics, which are necessary for visualization, while procedural models will be aligned to every possible operational scenario. Through the SysML approach, we divide the system model into two parts: the structural model and the procedural model. The structural model consists of the Block Definition Diagram (BDD) and the Internal

Block Definition Diagram (IBD). The IBD shows the internal structure of the parts and the interaction points (SysML ports) of the model with the external world. The procedural model shows a diagram of the system that represents the different states the model can be in, for example, "maintenance" (state 1) and "operational" (state 2). The change of state requires a transition that can be an internal or external event, a signal or a time event. Each state of the model represented in the state diagram can have one or more activities that describe the behavior of the model in that state. An activity diagram (ACT) models the flow of actions or events that must be performed to complete a procedural function.

In this way, an independent SysML model can be built with ports for interconnection with other models. Models are interconnected by means of ports. Ports can model the flow of signals from one model to another, but also require operations, an executed action, or a flow of materials. Behavior model state machines can contain transitions that depend on a transition or behavior occurring in another model and thus another model, thus allowing interaction between models. In this way, SysML can model the actor, product, and environment separately.

All geometric and procedural model information in the scenario under consideration is loaded. The SysML model must naturally contain the description of the scenario that will be loaded into the virtual environment so as to allow the user to act and possibly modify requirements and geometric properties of the main model. At this point, the systems engineer observes the scenario within the XR; he observes the user acting with/on the main product (aircraft under maintenance), the reaction to actions, and the behavior of the environment.

The system engineer must clearly load or suppress some elements of the system and environment if they believe that the requirements and even the goals have changed. This mechanism makes the (re)configuration of a VR scene easier and faster than the current way of working where, after each change, the whole scenario has to be rebuilt.

CHAPTER 3   XR LIFE CYCLE

In general, we are stating that the systems engineer needs to interact and create multiple scenarios and to verify, for example, product usage. Since models built in SysML have high reusability, they can be used in comparable scenes or even for the next one.

Another interesting example would be related to performance improvement in a production line. The approach used is definitely that of Lean Manufacturing. First, this approach is a production management method based on waste elimination and just-in-time production. This leads manufacturers to update and adapt their production management methods and reorganize production lines to reduce costs, minimize delays, and optimize flexibility. Although lean thinking has proven effective in helping professionals identify and eliminate waste during engineering operations, systematic instruction mechanisms and training protocols based on individual trainee performance are insufficient in existing training to define value-added activities for further productivity improvement in a training environment (see interview with Francis Vu in Chapter 7). How could this approach be further improved? If we consider Virtual Reality as a totally computer-generated environment, where the user using it is isolated from the real world, then its use would bring many benefits in terms of preventive simulation of production lines. In fact, VR would allow users to interact with the simulation model in real time and in 3D, proposing multiple methods of simulating production flows in a real-time virtual environment. Starting from traditional industrial tools in the field of manufacturing simulation, it is possible to derive which ones can be well represented by a virtual environment. Finally, a comprehensive software solution integrating VR will be proposed and an example of its application to simulate lean line production in VR will be presented.

Verification and Validation of XR

In the framework of defining and managing a system, verification and validation processes are a pillar of the SE approach. In this section, we will attempt to set out unambiguously how these processes are applicable to VR technology, which can be both a verification and validation tool and itself a digital technology to be qualified (see Chapter 4).

Verification

Verifying a system in the metaverse entails confirming the fidelity of the XR environment to its real-world counterpart. However, this alone is not sufficient. Validation, on the other hand, involves assessing whether the VR experience successfully achieves its intended objectives. Let us explain further. Suppose we want to measure the length of a table using a classic and rudimentary measurement tool. Now let us repeat the same operation in virtual reality. Well, if the two measurements are equal, bearing an acceptable error, we have uniquely verified the goodness of the virtual tool and the goodness of the measurement made through a digital tool. Therefore, for subsequent table measurements, we are confident that we can use the virtual reality that we have constructed for ourselves. It is clear and obvious that we also need to verify and validate the correct transposition of the table in 3D format and perhaps the eventual operator in the virtual environment taking a virtual gauge and measuring a virtual table. Of course, without belaboring the point too much, you understand well that the verification and validation of a VR must involve a whole series of analyses and tests designed to confirm that the observed reality is the same as the virtual one.

Before dealing with possible verification and validation processes using VR, let us make a small philosophical passage that serves to clarify what is meant by verification and validation. Although these terms may take on similar meanings in everyday usage that are outside the

CHAPTER 3 XR LIFE CYCLE

scope of technical scientific texts, for activities related to the field of systems engineering, their use is certainly specific and not arbitrary. First, let us clarify that verification assesses the degree to which simulations performed to define the goodness of a physical (conceptual) model are correct, while validation assesses the degree to which the mentioned physical model captures "reality" through comparison with experimentation or at most experimental measurements. Therefore, as we will delve into the scientific field, it can be safely said that verification and validation (V&V) can be seen as a logical extension of the standard scientific method. In this regard, clear and distinct processes need to be defined in order to emphasize a more rigorous, systematic, and quantitative approach and then accordingly verify the proposed models.

In a more formal sense, verification refers to confirming that a system or component meets specified requirements and functions as intended. On the other hand, validation entails demonstrating that a system or component satisfies its intended purpose in its intended environment. When it comes to VR technology, verification and validation are essential steps in ensuring that virtual experiences truly mirror reality. By subjecting VR systems to rigorous verification processes, developers can ascertain their precision and adherence to defined standards. Validation further ensures that these VR experiences accurately represent real-world scenarios.

Verification can be traced back to a "simple" applied mathematics and logic problem, although it usually does not have a rigorous solution, rather we rely on extensive testing to prove that the simulation is "correct." Clearly, since the model does not provide expected outcomes, it is necessary to evaluate the errors in the simulation. There are two general sources of error that must be addressed by verification:

- Errors in the "code/algorithm"
- Errors in the computational model solution

If the first errors can be easily tied to misunderstanding (the behavior of the real system) and then numerical transcription of the conceptual model, errors found in the solution may stem from the method used to derive it. This means that the method used to find a solution may not converge, or it may not evaluate any disturbances, boundary, or initial conditions.

Clearly, if verification is performed by a computer, the question inevitably arises as to whether the continuous mathematics of the model is correctly solved by the discrete mathematics of the computer code. Therefore, the first step is to ensure the quality of the created ad hoc software throughout the development process. Thus, the goal is to reduce errors as much as possible. It then turns out that software is in continuous development, so design and maintainability of the software are key requirements.

The ideal tool for verifying solutions is to compare simulations performed either analytically or on benchmarks. Code benchmarking should not be viewed as a substitute for a baseline analysis.

Validation

Validation is essentially a physical problem, but it should not be seen as a single, one-time operation in which codes are accepted (1) or rejected/discarded (0). Therefore, it must be understood that validation is a cyclic activity in which models are continuously improved.

At this point, we are able to abandon the philosophical approach and can express verification and validation as follows: for a defined set of problems and applications, a relative set of variables with a given level of accuracy and precision are assumed. The goal of validation and verification is an assessment of the extent to which a simulation represents the actual behavior of the system to a sufficient degree to be useful.

CHAPTER 3 XR LIFE CYCLE

How can we translate this into VR? And what fundamental parameter derivation processes can we accept in accordance with the real observed model? Meanwhile, for VR, the basic parameters are defined because they can be measured and users can see how they vary when imposed as results of a simulation. Therefore it would seem logical to create several models, different from each other, that deal with the same basic parameters considered. After that, we compare the simulations with different measurements at different levels of the established hierarchy based on the level of parameters/measures deemed basic or derived from other quantities. For example, in the VR representation of an aircraft maintenance environment, the lowest level of the hierarchy may be: the failures reported directly by the on-board computer (e.g., hydraulic system failure), while at the next levels: the derived quantities (e.g., pump inoperative, fluid leakage etc.). In general, the discrimination between models decreases as we go up the hierarchy, although comparison at multiple levels of the hierarchy is the best practice. Note that the shape of the hierarchy is not necessarily unambiguous: the important thing is to learn how to deal with uncertainties and errors.

Before the advent of VR, the management of uncertainties and errors had been left to synthetic diagnostics that attempt to numerically replicate, in all its details, the physical processes of geometry, spatial and temporal visualization that characterize the actual measurement. The development of any synthetic diagnostic is essentially a complex exercise in matching phase-space geometries that requires careful characterization of the physical diagnostic. The synthetic diagnostic code can be quite complex and must be thoroughly tested. As mentioned above, comparison generally provides some of the most fundamental validation tests. Clearly, mathematics constantly helps us through a wide range of tools such as, for example, self-power or spectral density functions, which are easy to compute even with limited amounts of data. Unfortunately, these are often limited when it comes to discriminating between interacting or complex because they are sums of numerous elements.

CHAPTER 3   XR LIFE CYCLE

In short, we understood that the V&V stage in a system life cycle is one of the most difficult activities to manage both in terms of cost/time and feasibility. V&V processes occur after the completion of detailed and specific work delegated to specialists to ensure that the entire system of interest conforms to project specifications and requirements. It is clear that the processes described above do not produce satisfactory results the first time around, so an iterative path is run based on errors, in a generic sense, found during verification. The systems engineer is responsible for recording information on the anomalies found from the comparison of criteria and requirements established by all stakeholders. It is equally clear that the systems engineer is not always able to determine how many times an iterative process of V&V is carried out and therefore difficult to determine precisely how much time is spent on the activity. Instead, let us see if VR can at least help to improve the said process, which, in fact, is currently tied to the ability of a person able to collect the necessary information by model comparison.

First, let's make the list of elements to be digitized in three-dimensional (3D) form:

- Project information and relevant data to be integrated, including planning and function diagrams
- Geo-location information and spatial data
- Information, if any, of the environment in which the system operates (e.g., for aviation maintenance, it may be necessary to include the logistics of the components to be maintained)

This approach, while supportive of the V&V phase of any constituted system, may have several limitations; we can enhance it through AR or VR, depending on the visualization needs and interactions with virtual content. Although AR technology seems to be a promising means to enhance communication and integration of subsystems, the effectiveness of this technology is not always proven. In fact, AR, especially in the manufacturing industry, has been applied significantly more for system development rather than for validation.

We try to better understand this approach by investigating the feasibility and practicality of AR technology when we wish to perform a certain maintenance. Let us try to understand whether it is possible to validate maintenance through VR and/or AR. Generally, the maintenance process is manual (e.g., checklists, drawings, procedures, instructions etc.) in which the technician performs and possibly records any anomalies found. The information must be passed on for possible subsequent action, such as replacing the damaged part once the "new" part is made available. Needless to reiterate, this process is time-consuming and depends on the maintainer's skills, training, and experience to identify defects.

First of all, the defect detection system based on the maintenance manual (provided not only in the aviation field but also in the civil field) can be integrated with mobile AR technology to process a series of images. In fact, as we have already mentioned in previous chapters, AR allows to anchor 3D simulations in space and seamlessly overlay them to reality. Such a system would enable remote inspection, for instance, it would allow the maintainer to detect dimensional errors and omissions automatically.

Surely now we are wondering what is the acceptable degree of accuracy. Well, this is the main challenge of VR. Meanwhile, we need to align the maintainer's AR camera with the virtual model. After that, we overlay the digital mode with the real environment using a wearable AR mobile device. We can now perform AR measurements. According to the V&V process, we need to compare the conventional (manual) maintenance method with the proposed system and demonstrate whether, for example, we have a minimal margin of error while gaining a significant improvement in time (to perform the task) as compared to conventional inspection practices. If the use of AR resulted to be time-neutral to the "real" activity, the AR approach would clearly be useless.

Let's get to the heart of the discussion. Let us take up the example shown in Chapter 2. The goal is to simulate an inspection through a VR environment if the operator is remotely located. Surely we will have the digital mockup of the aircraft and its subsystems (e.g., architectural, mechanical, electrical) that need to be integrated into a dedicated platform. Theoretically, several AR devices can be used, with the digital content (latest configuration or the model) appearing on the screen of the device and allow us to move freely in the virtual space, possibly enhanced with hand-gesture interaction.

Therefore, the implementation of this system in maintenance must consider:

- The tracking system. This could be marker-based, so it requires setting a physical marker on the aircraft that exactly matches that of the digital version.

- Run the application and detect the marker using an HMD camera by using the appropriate image recognition algorithm.

- Once the marker is recognized, the application starts retrieving data on the aircraft and overlaying it on the existing structure.

From there, the maintainer can check troubleshooting activities and compare them with the designed model without needing to go from the physical environment to the drawings and back again. Aircraft components and their properties can already be stored in the XR system, the user can choose any available component in the displayed scene using gaze and finger tap to retrieve them. It is worth mentioning that in situ virtual overlay is an essential factor in AR applications, as it is required to bring the geometric data to full scale and precisely fit it to the existing structure through the virtual overlay enabled by image recognition techniques (marker or markerless). Therefore, the marker must be placed at an appropriate location in both the real and virtual environments, which share the same coordinates. If we want to conduct a visual inspection on the fuselage of an aircraft, the physical image-marker must be placed on a surface finish layer to achieve an accurate alignment of the holographic data. If, on the other hand, we wanted to troubleshoot in the cockpit, the physical marker must be in a condition of excellent illumination since VR hardware must always recognize the image-marker. In general, the marker-based approach is a preferred solution when high accuracy is required and the maintenance environment is free of external disturbances, especially if it comes to be geographically located in remote areas.

Digital Continuity

Digital continuity is a crucial aspect of the architecture life cycle, ensuring seamless integration and consistency throughout the entire process. One effective way to maintain digital continuity is through the implementation of a structured breakdown and parametric modeling approach. By breaking down the architectural design into structured components, it becomes easier to manage and track changes at each stage of development. This approach allows for greater control over modifications, ensuring that any alterations made are accurately reflected across all relevant aspects of the project. Furthermore, parametric modeling provides a powerful tool for maintaining digital continuity by establishing relationships between different elements within the design (see Interview with Paul Davies in Chapter 7). These relationships allow for automatic updates whenever adjustments are made to one component, ensuring that all interconnected parts remain synchronized. It is clear that the XR simulation needs to be part of this digital continuity. The combination of structured breakdown and parametric modeling not only enhances efficiency but also promotes accuracy in different phases of the system life cycle. By keeping digital continuity intact throughout the entire life cycle, engineers can minimize errors and inconsistencies while maximizing collaboration among team members. In conclusion, embracing a structured breakdown and parametric modeling approach is essential for preserving digital continuity, an essential building block of the industrial metaverse (see interview with Dr Laughlin in Chapter 7).

Takeouts

- Since the late 1980s, design activities for generic products have been conducted concurrently to produce high-quality performance products while minimizing costs and time-to-market. This approach is aided by rapidly evolving information technology, such as Computer Aided Design (CAD) support and computer simulations. VR technology aims to solve the limitations of current computer tools, from simple visualization of 3D models to dynamic and interactive representation of the product. This chapter provides a comprehensive explanation of how a generic product can seamlessly interact within a dedicated environment, considering all aspects of the life cycle. XR product development can help meet industrial requirements related to cost, quality, and time. By integrating VR into the system life cycle, it facilitates activities contemplated in each stage, integrating methodologies from SE literature. MBSE, a model-based approach, can introduce all the information needed to define the life cycle of a system into a VR environment, allowing for better collaboration, visualization, and decision-making throughout the product development life cycle.

- XR applications in a system life cycle can be improved by incorporating procedural descriptions and allowing users to manipulate models in a given environment. Currently, there is no acknowledged method for building procedural models in XR, leading to the need for a new, efficient method for proper configuration. A virtual environment must provide the correct representation of the scene, reusable models, and extended user interaction. SysML can be used as the language for procedural descriptions in XR, dividing the system model into two parts: the structural model and the procedural model. The system engineer can load or suppress elements of the system and environment if requirements or goals change, making re-configuration easier and faster. Virtual Reality (VR) can also be used for performance improvement in production lines, allowing users to interact with simulation models in real-time and 3D. A comprehensive software solution integrating VR will be proposed, demonstrating its application in simulated lean line production.

- Verification and validation are crucial in the field of systems engineering (SE) to ensure that virtual reality (VR) experiences accurately mirror reality. In the metaverse, verification confirms the fidelity of the XR environment to its real-world counterpart, while validation assesses whether the VR experience achieves its intended objectives. These processes involve a series of analyses and tests to confirm that the observed reality is the same as the virtual one. Verification can be traced back to applied mathematics and logic problems, while validation addresses errors in the simulation. The ideal tool for verifying solutions is to compare simulations performed analytically or on benchmarks.

- Validation is a cyclic process that involves continuously improving models to assess the accuracy and precision of a simulation. In Virtual Reality (VR), basic parameters are defined and compared with different measurements at different levels of the hierarchy. This helps manage uncertainties and errors, which are difficult to manage in terms of cost/time and feasibility. The V&V stage in a system life cycle is challenging to manage, as iterative paths are run based on errors found during verification. VR can help improve this process by improving the ability to collect necessary information through model comparison.

- The text explores the use of 3D digitization in aviation maintenance, focusing on project information, geo-location information, and environmental information. The use of AR or VR technology can enhance communication and integration of subsystems, but its effectiveness is not always proven. The text investigates the feasibility and practicality of AR technology in maintenance, focusing on remote inspections. The main challenge of VR is determining the acceptable degree of accuracy. The implementation of this system in maintenance requires a tracking system, which can be marker-based or HMD-based. The application retrieves data on the aircraft and overlays it on the existing structure, allowing the maintainer to compare troubleshooting activities with the designed model.

The use of a structured breakdown and parametric modeling approach is essential for maintaining digital continuity throughout the system life cycle. This approach minimizes errors and inconsistencies while maximizing collaboration among team members.

Questions

- How does Virtual Reality (VR) technology enhance the design process?

- What are the advantages of using SysML as the language for procedural description in virtual environments?

- How do verification and validation processes differ in the metaverse context compared to traditional systems engineering?

- Why are uncertainties and errors difficult to manage in terms of cost, time, and feasibility in VR simulations?

CHAPTER 4

Convergence

Sometimes, the immersive industry feels like you're in the back of a car while the driver keeps saying 'we're almost there' for about ten years. The end is just around the corner, or over the next hill, or by the town ahead. But never is it here, now, in the present. Metaverse discussions are even worse, where you are unsure where the driver is even going.

—Ffiske, Tom. The Metaverse: A Professional Guide : An expert's guide to virtual reality (VR), augmented reality (AR), and immersive technologies (p. 87). Kindle Edition.

This chapter introduces the concept of convergence in Extended Reality (XR) frameworks within the context of Model-Based Systems Engineering (MBSE). This chapter refers to the integration and alignment of various technologies, methodologies, and tools to create cohesive and immersive experiences. The convergence is driven by the need to enhance collaborative systems engineering processes, improve decision-making, and facilitate stakeholder engagement in complex system development. The convergence transforms the systems engineering landscape by providing advanced tools that facilitate collaboration, visualization, and decision-making. This integration leads to more effective complex system design and development, ultimately resulting in high-quality outcomes that meet stakeholder expectations.

CHAPTER 4   CONVERGENCE

Convergence Primer

The term convergence, in the context of the Industrial Metaverse, means using a single interface that manages all the systems and subsystems of the SoS. Such an interface will be necessary to achieve the SoS objective. Convergence has been established in the system model, indicating that the primary variables will approach their equilibrium levels. As a result, the other variables in the system should also move towards their respective equilibrium points.

As we saw in the previous chapters, extended reality can play a significant role in developing a truly model-centric system engineering approach. Furthermore, the core application of the MBSE approach implies the use of a single system model that serves as the "source of truth" for all embedded system analysis tools. A system model can therefore interact with a virtual reality environment to analyze system behavior. In this framework, the role of systems engineers can be introduced because they can run interactive, immersive simulations of scenarios described in system models and feed results from virtual reality environments back into system models. Based on the results of the interaction, system engineers can change product/service specifications and designs. In fact, by utilizing XR environments, the engineering analysis can be performed more efficiently by involving the customer in the product life cycle from an early stage (see Chapter 3).

The terminology XR4MBSE and MBSE4XR, mentioned for the first time in this book, is an extrapolation of the popular, among systems engineers in the early 2020s, reference to AI4SE and SE4AI (see source "A. Migliaccio, G. Iannone (2023). Systems Engineering Neural Networks, Wiley." in Appendix B). This refers to the binomial convergence of two disciplines, where the first is used in support of the latter and vice versa. In this case, those acronyms are intended to convey the following.

XR4MBSE is a term that stands for extended reality applied to model-based systems engineering practices. This includes the usage of virtual simulations and augmented reality to aid engineers in their

CHAPTER 4   CONVERGENCE

tasks, such as verification and testing at multiple stages of the process. Also, with XR4MBSE, engineers can make use of virtualizing CONOPS (concepts of operations) and enhanced visualization capabilities to ensure accuracy and save time when engineering models. XR is becoming increasingly popular among engineers for its ability to improve accuracy, efficiency, and cost savings for engineering operations. It offers several benefits such as

- Creating a shared space for discussing designs with stakeholders from remote places in real time

- Designing complex scenarios more quickly from scratch or expanded versions from previous designs

- Testing results during development cycles through virtual simulations, reducing physical burden on system components

MBSE4XR, the use of MBSE to enhance XR technology, is related to all the modeling techniques used to develop XR systems (hardware, software, language, interfaces, etc.). Therefore XR is intended as a complex system to be analyzed. In fact, with the advent of extended reality, systems engineering is proving to be one of the most systematic processes to design immersive experiences that engage users in a way unlike ever before. While the concept of using "systems'" thinking to enhance XR experience design has been around for some time, its application is becoming more popular thanks to Model-Based Systems Engineering (MBSE4XR) and other model engineering techniques.

Note   The authors would like to stress that this MBSE4XR approach is best covered in other books dealing primarily with the systems methodology (see Sources in Appendix B).

CHAPTER 4   CONVERGENCE

Implementing an XR framework requires careful consideration of various factors. Here are some that are recommended:

- **Research and Analysis**: Thoroughly investigate existing norms, regulations, and safety guidelines pertaining to XR technology in industrial settings.
- **Stakeholder Engagement**: Engage with relevant stakeholders such as regulatory bodies, industry experts, employees, and management to gain insights into their needs and concerns.
- **Risk Assessment**: Conduct a comprehensive risk assessment to identify potential hazards associated with XR implementation in different industrial processes.
- **Customization**: Tailor XR solutions based on specific industry requirements by collaborating closely with production line managers or supervisors who understand operational nuances.
- **Training Programs**: Develop training programs that address both technical aspects of using XR technology as well as safety protocols related to its application in industrial environments (more details in Chapter 6).
- **User Experience Testing**: Prioritize user satisfaction by conducting extensive testing sessions with workers involved in various roles within the production line to gauge effectiveness and ease of use.
- **Compliance Monitoring**: Establish mechanisms for continuous monitoring of compliance with established norms and regulations throughout the implementation process.

CHAPTER 4　CONVERGENCE

- **Data Security Measures**: Ensure robust data security measures are implemented to protect sensitive information gathered through XR systems used within industrial processes.

- **Scaling up Possibilities**: Explore opportunities for scaling up successful implementations across different units or departments within an organization or even across multiple organizations within an industry.

- **Industry Collaboration**: Encourage collaboration between industry players, researchers, and regulatory bodies to share best practices, address emerging challenges, and foster innovation within the XR development space.

The Importance of Continuity and Life Cycle Mindset

Described as a "key enabler" for smart connected products, new business models, financial transparency, and faster time to market, **digital continuity** helps avoid information silos and enhance industrial collaboration. This allows teams, departments, or the entire enterprise to coordinate activities better and increase productivity. The next chapters delve into the various key enablers that contribute to the success of creating a metaverse. As shown in the Figure 4-1, the enablers include the infrastructure and data distribution system, the interfaces of devices within a company, and the availability of XR applications – whether they are self-generated or obtained externally through third parties. Infrastructure plays a crucial role in establishing a solid foundation for the metaverse. This includes aspects such as bandwidth and servers, which are essential for smooth data transmission and processing. A well-designed infrastructure ensures that digital assets can be accessed seamlessly and enables digital continuity across different platforms.

Furthermore, the interfaces of devices within a company are vital for achieving interoperability (see interview with Paul Davies in Chapter 7). The ability to seamlessly connect various devices allows for a cohesive user experience in the metaverse. Efficient communication between these interfaces ensures that users can interact with digital assets effortlessly. Finally, one must consider both self-generated and purchased XR applications when building a metaverse ecosystem. Whether an organization chooses to create their own applications or acquire them externally (make/buy), having access to an array of XR applications is crucial for creating immersive experiences within the metaverse.

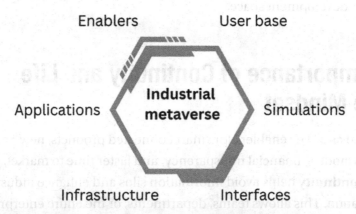

Figure 4-1. *Key enablers for Metaverse implementation*

Digital continuity and life cycle (Figure 4-2) mindset play a key role in the usage of the Industrial Metaverse. For example, we are surrounded by systems that have been designed by performing simulations: structural, functional, thermal, etc. It is safe to say that in most cases, such a quantity of generated data remains confined to the purpose it was meant to serve, without moving across to other activities covered in the industrial domain.

In fact, a possible misalignment between different departments in the same company could be commonplace (silo mindset). For example, during the dismantling of a car, the department dedicated to recycling does not have access to the installation drawings of the different parts that are probably only available within the design department.

Figure 4-2. *Product Life Cycle Management*

During the development of a generic product, it is important that the concepts of digital continuity are introduced, the reuse of data can support PLM (Product Lifecycle Management), along with visualization components, exchange of information between processes, and suitable software interfaces. PLM is a systematic approach to managing the sequence of changes that a product undergoes, from design and development to final retirement or disposal. PLM is typically divided into the following phases:

- Beginning of Life (BOL) – Includes the new product development and design process.

- Middle of Life (MOL) – Includes supplier collaboration, product information management (PIM), and warranty management.

- End of Life (EOL) – Includes strategies for how products are retired, discontinued, or recycled.

PLM software applications, which may or may not include XR components, help companies manage the life cycle of their products by providing a **data lake of all product-related information**. PLM software can automate the management of product-related data and integrate the data with other business processes such as enterprise resource planning (ERP) and manufacturing execution systems (MES). The goal of PLM is to eliminate waste and improve efficiency. PLM is considered an integral part of the lean manufacturing model.

Emerging Disciplines

In Chapter 3, we have seen the utilization of extended reality to support different phases of the system life cycle, although it is key to define the disciplines that are involved in the different phases. Some of these are not yet fully defined, as model-based engineering techniques are still being codified. Although some definition can already be drawn:

Digital Engineering: An integrated digital approach that uses authoritative sources of systems' data and models as a continuum across disciplines to support life cycle activities from concept through disposal. (DAU Glossary) (Defense Acquisition Guidebook)

Model-Based Engineering: An approach to product development, manufacturing, and life cycle support that uses a digital model and simulation to drive first-time quality and reliability. (NIST)

Model-Based Systems Engineering: Execution of Discipline (Systems Engineering) using digital model principles for system-level modeling and simulation of physical and operational behavior throughout the system life cycle. (INCOSE)

Architecture: System fundamental concepts or properties of a system in its environment embodied in its elements, relationships, and in the principles of its design and evolution. (ISO 42010)

CHAPTER 4   CONVERGENCE

These disciplines are interesting to have a look at what professional figures are going to be essential to this phase of digital transformation. In fact, at the time of writing, a consolidated discipline, that we could call MBSE Virtual Engineering, does not yet exist. Anyway, they are all essential when thinking of infusing or designing an organization which is MBSE-savvy and wants to incorporate immersive technology seamlessly in its processes, the same way that simulations are now part of most, if not all, productive processes that require manufacturing.

Current and Speculative Professional Roles

We could wonder whether MBSE is a competence or a role? MBSE is a methodology; in fact, at the time of writing, the professional community is divided on this point, therefore, a reasonable extrapolation from current professional figures can be made to better understand the players involved and to help the reader understand how to create a suitable team to tackle the challenges of XR-based virtual engineering.

Global or Local MBSE Architect

A systems architect specializes in designing and developing computer networks and systems for company operations. Typically, their responsibilities revolve around conducting research and analysis to identify a company's needs, devising strategies to reach particular goals, improving existing systems, and implementing solutions for optimal processes. A systems architect must also monitor the progress of all implementations, prepare progress reports and presentations, provide technical support as needed, and train new workforce members, all while adhering to the company's policies and regulations, including vision and mission. This role is necessary whether or not immersive technology is included in the system, although it is recommended for the system architect to be assisted by a subject matter expert.

MBSE Engineer with XR Expertise

An MBSE engineer working in the IT Agile/Waterfall environment is accountable for the successful technical analysis, planning, architecture, and design of the solution, implementation, and tests up to validation of the Engineering tools/software projects, with monitoring, control, and support to their execution.

XR/Immersive Producer

A XR Producer is responsible for managing the programmers, developers, and day-to-day processes of all projects, and will work with senior Creative and Technical teams to define development, production budget, and timelines for all productions.

Virtual or Digital Engineer

Virtual Engineer provides a user-centric, first-person perspective that allows users to interact naturally with the system built, giving users a wide range of accessible tools. This requires an engineering model containing geometry, physics, and quantitative or qualitative data of the actual system. The user should be able to step through the operating system and observe how it behaves and how it responds to changes in design, operation, or other engineering changes. Interactions within the virtual environment should present an intuitive interface that is appropriate to the user's technical background and experience. This allows users to explore and discover unexpected but important details about system behavior.

MBSE-XR Champion

Implementing Model-Based Systems Engineering (MBSE) within an Extended Reality (XR) framework can be a transformative process

CHAPTER 4 CONVERGENCE

for organizations looking to enhance their design and development capabilities. The implementation process (Figure 4-3) tailored for an MBSE-XR Champion provides a robust approach that leverages the strengths of both MBSE and XR, ultimately enhancing product development and system design processes. This implementation creates a more collaborative and interactive environment that can lead to better design outcomes and a more effective engineering workflow. They can give feedback to the teams in charge of improving immersive experiences and procedures. They are active users of any means supporting the implementation of use cases for the Industrial Metaverse. The process to set up a champion network in your organization will be covered in more detail in the next chapters.

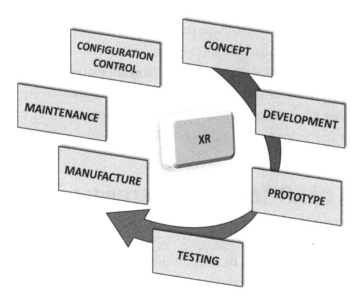

Figure 4-3. *XR as product – implementation process*

CHAPTER 4   CONVERGENCE

Narrative Designer

At first, the term "narrative" in the Industrial Metaverse refers to the design of dynamic experiences in an industrial framework. Therefore, a narrative designer plays a crucial role in the Industrial Metaverse development and utilization. The focus of a narrative designer is to design the narrative elements based on how users interact with its "story."

This role combines elements of design and engineering to create immersive and engaging experiences for users. Narrative designers work closely with other members of the development team to ensure that the story and engineering simulations are seamlessly integrated.

One key aspect of being a narrative designer is understanding how to enhance the user experience by creating dynamic narratives that respond to user's choices and actions, these designers can create experiences tailored to the needs of each engineering team.

In summary, narrative designers bring together storytelling, design, engineering, simulations, and model-based systems to craft compelling narratives that adapt to user interactions. Their role is essential in shaping the overall experience and immersion within the Industrial Metaverse.

Impact of XR-MBSE Convergence on Industrial Activities

In the previous chapters, we have mentioned that XR deployment in an industrial environment, without a solid modeling base, is not easy to implement and can severely limit its advantages.

Solid modeling involves more than just graphical elements as it encompasses evaluations related to the generation of complex containers that converge the main elements of a system life cycle. The literature suggests that solid modeling is an essential aspect in the development of complex systems.

These are further exemplified by these quotes:

- "Most efforts to define new graphical conceptual modelling notations are spent on designing semantics while ignoring or undervaluing the role of visual syntax" [Moody, 2009].
- "Practitioners spend inordinate effort transposing models from rigorous tools to non-structured formats to overcome the acceptance problem. In the process, the connection to the source repository is lost, thereby destroying integrity, reusability, maintainability and currency of the derived output when the source changes" [McLeod, 2018].
- "The poor quality of graphical conceptual modelling notations harms the acceptance from the non-specialist audience" [McLeod 2018; Moody, 2009].

Effective modeling should incorporate life cycle management to quantify and analyze the data, information, and processes involved in creating a product or service, including the associated costs and time constraints. Product Lifecycle Management (PLM) enables this by providing a shared platform for managing the product's entire life cycle, from concept to disposal, allowing organizations to identify and mitigate any negative impacts throughout the process.

PLM integration can result in the loss of vital information for the final product phase due to the lack of digital linkages between main life cycle phases. This disconnect can lead to inefficiencies and errors at the end of the production phase. This disconnect can be easily identified during the operational phase, causing a hindrance to the product performances. The disconnects between life cycle phases are a common issue that needs to be addressed to ensure the smooth functioning and success of a product.

CHAPTER 4 CONVERGENCE

It is essential to bridge the gap between life cycle phases to ensure that valuable data is shared and utilized throughout the entire product development process.

Current engineering processes for designing industrial systems employ a model-based paradigm. Limitations on developing multifaceted models (Zeigler et al. 2018) and lack of continuity in PLM have left the potential for collaboration between engineering, manufacturing, and support processes underutilized. With respect to digital continuity and interoperability, certain shortcomings of industrial system design have been identified as follows:

- Model-based systems engineering (MBSE) approach aims to ameliorate the above shortcomings through a unified system description using models.
- Different modeling languages for behavioral models in different domains; traceability of manual requirements.

Deep modeling in a virtual environment has the potential to address limitations in PLM. This approach can simulate various activities, including system requirements, conceptual design, analysis, V&V, design, and production (Figure 4-4). By leveraging deep modeling techniques, companies can better understand how products will perform under different circumstances, identify potential issues early on, and make more informed decisions throughout the development process. Ultimately, this can lead to more efficient and effective product development, as well as improved product quality and customer satisfaction.

The integration of Extended Reality (XR) into a model-based systems thinking approach and throughout the product life cycle presents a myriad of benefits. To make this information more accessible to the reader, these advantages/benefits are classified into four distinct, yet logical categories.

CHAPTER 4 CONVERGENCE

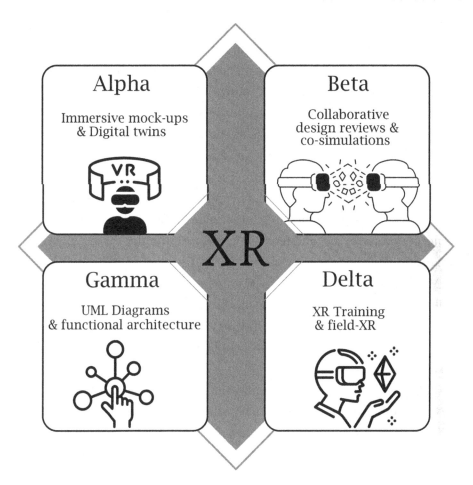

Figure 4-4. Illustration of the four groups proposed for XR implementation

Group Alpha – Immersive Mock-ups and Digital Twins

Digital twins have become an integral part of many industries, from automotive to aviation. These digital models provide real-time data about a product or system and can help monitor, analyze, predict, and optimize

CHAPTER 4   CONVERGENCE

performance. A digital twin relies on a 3D model to accurately reflect the physical state of the product or system being monitored. This is just one part of creating a digital twin – advanced techniques such as Model-Based System Engineering (MBSE) and Product Lifecycle Management (PLM) are also key to achieving digital continuity across applications.

In a nutshell, digital twinning is the process of creating a highly realistic model of a device, system, process or product to use for development, testing and validation. Augmented reality (AR) and virtual reality (VR) come into play as well. For example, AR can show a digital twin on top of a physical machine and provide information a technician wouldn't otherwise see, and technologists can enter the VR of a digital twin to simulate various issues.

Many associate the use of digital twins solely with manufacturing. While it is true that manufacturing has pioneered the use of digital twinning, use cases exist in every industry. Additionally, there are digital twinning use cases in cross-industry applications such as infrastructure and automation.

To better understand the potential uses of digital twinning, AR and VR, take a look at the use cases in a handful of industries. Aerospace, automotive, and general-purpose manufacturing firms use digital twins as part of overall product development. Here are some common uses:

- Creating design mock-ups to show how a finished product will work
- Fine-tuning product features and capabilities
- Defining requirements to provide guidance to component suppliers on component specifications, such as bolt size, shape, and strength
- Testing and quality assurance
- Creating customer-requested modifications and other design personalization

- Creating operational and performance optimization and

- Predicting future failure modes so maintenance can be preemptively scheduled, and executing on other predictive maintenance goals

In addition to the general advantages of using immersive technology in combination with digital twins, two aspects of how immersive technology in combination with MBSE can enhance industrial processes are presented as follows: Zero Defect Manufacturing and Automatic Simulations.

Zero Defect Manufacturing System

How can XR limit errors and information loss within a PLM? At the end of the 1960s, a Zero Defect (ZD) program was developed in American industry to eliminate most defects in production. In general, the ZD theory guarantees that there is no waste in a project and definitively eliminates all those unproductive processes, including instrumentation. Above all, this theory expresses the fundamental concept of "doing it right the first time" to avoid expensive and time-consuming solutions.

Furthermore, Spare Manufacturing (LM), Six Sigma (SS), Theory of Constraints (TOC), Total Quality Management (TQM), and Lean Six Sigma (L6S) are examples of conventional quality improvement methodologies. These Quality Improvement (QI) methods have the objective of enhancing product quality by identifying and eliminating defects, rather than learning from past mistakes. They do not fully utilize advanced data-driven technologies as assessed by the concept of Industry 4.0 (see Chapter 1). Moreover, these methods do not take into account the prediction or consequences associated with it. While these strategies may provide some support during the digital and green transition, they can only offer a limited response and assistance to address new challenges that arise in this context.

CHAPTER 4   CONVERGENCE

Zero Defect Manufacturing (ZDM) is a modern strategy that aims to address the limitations of traditional Quality Improvement (QI) methods. This approach leverages digital technologies such as Artificial Intelligence (AI), Machine Learning (ML), and large artificial datasets to intelligently anticipate and prevent issues in product and process scenarios. By doing so, ZDM enhances autonomy and improves overall quality control. It leverages the ability to incorporate predictive outcomes and facilitates the implementation of comprehensive feedforward and feedback control loops. Both of these elements work together to effectively accomplish adaptable and sustainable objectives.

In addition to the strategies employed by the ZD philosophy for evaluating waste, implementing process changes and promoting staff proactivity while monitoring progress, the introduction of Extended Reality (XR) technology brings a new dimension to enhancing process performance. XR has the capabilities to simulate various processes and provide continuous feedback on activities carried out. Through XR simulations, organizations can gain valuable insights into potential improvements and identify areas of inefficiency or waste. Furthermore, it enables continuous feedback loops that allow for immediate adjustments and improvements based on real-time data.

Overall, incorporating XR into PLM workflows can help streamline the development process and improve product quality. In manufacturing, quality control measures should be put in place to address issues, such as regular inspections, testing, and employee training. XR applications for supporting simulation of assembly processes is the use of Augmented Reality (AR) to provide step-by-step guidance and visual feedback to assembly workers. This can improve accuracy and efficiency, reducing errors and eliminating defects. Additionally, workers can make practice before performing the production tasks. This can enhance safety and reduce the risk of damage to components or equipment. Finally, in the manufacturing phase, XR applications can provide continuous improvement which will be shared in real time with the development team.

CHAPTER 4   CONVERGENCE

Another example of an XR application for supporting logistic warehouse staff is shown in Figure 4-5. XR technologies offer a wide range of opportunities for logistics support that can help save time and reduce costs. Overall, XR can be used to simulate real-life scenarios and assist with logistics planning that can be optimized by collaboration with vendors and carriers. Additionally, XR applications such as Mixed Reality (MR) can be used to visualize data and track inventory levels in real time, helping logistics providers to optimize their supply chains.

Figure 4-5. *Example of Augmented Logistics*

Automatic Simulations

The integration of artificial intelligence and virtual reality has paved the way for a new concept known as automatic simulation. This cutting-edge technology enables the simulation of various scenarios and visualizes product performance in the metaverse, all in an automated manner.

CHAPTER 4   CONVERGENCE

By leveraging artificial intelligence algorithms, automatic simulation allows organizations to create immersive experiences that go beyond traditional methods. Scenarios can be created and manipulated with ease, providing insights into how a product will perform under different conditions. By simulating different scenarios, you can assess the product's strengths and weaknesses, identify potential challenges, and make informed decisions. Scenario analysis allows you to explore the impact of different variables on your product's performance. For example, you can test how it responds to changes in customer demand, market trends, pricing strategies, or even external factors such as weather conditions. Moreover, scenario analysis enables you to evaluate the effectiveness of contingency plans or alternative strategies. By simulating various scenarios and their outcomes beforehand, you can make well-informed decisions that minimize risks and maximize success.

In the past, subsystems required for simulations were often siloed in different databases and accessible by other specialists. However, with digital continuity, these subsystems are now integrated into a unified metaverse, enabling seamless access and collaboration.

The potential benefits of automatic simulation are vast. Organizations can quickly identify design flaws (see interview of Paul Davies in Chapter 7), evaluate variations, and make informed decisions before investing in physical production. This technology also opens up opportunities for cross-reality (XR) experiences where users can interact with products virtually.

With automatic simulation, companies can harness the power of artificial intelligence and virtual reality to enhance their product development processes. By automating simulations in the metaverse, they can gain valuable insights into product performance while saving time and resources.

CHAPTER 4   CONVERGENCE

Group Beta – Collaborative Design Reviews and Co-simulations

Over the past few decades, Collaborative Design Reviews (CDRs) have made use of cave immersive systems for a more collaborative experience. Although these systems do not normally enable designers and engineers to share 3D CAD models, data-driven diagrams, MBSE models, PLM data, and legacy documents in real time in one place. As organizations shift to digital transformation and digital continuity strategies, they are increasingly using digital twins for their CDRs. By leveraging 3D models associated with product data on a single platform, stakeholders involved can easily participate in review meetings with an immersive environment from anywhere in the world – making the process accurate at scale and saving time for design and engineering teams.

One necessary techno brick for the use of immersive technology in design reviews is the digital continuity, or digital availability of data and simulations that are different in nature or made for different purposes. Digital Continuity consists of three academic areas: data, simulations, and models. The experimental methods used in each of these are different. Data is gathered with a device that collects information from the world around us, which can be easily and quickly measured by a computer or other device such as an oscilloscope or an instrument. Simulations are created by using computer-generated programming to model complex systems like, for example: weather patterns, power grids, wars, and economies to help understand how they function. Models are a key component of the continuity that we already discussed in Chapters 2 and 3.

Validation Through XR Simulations

As discussed in Chapter 3, verification and validation (V&V) is a fundamental part of any systems engineering process. It involves testing the system in various conditions to ensure that it meets its design specifications. As many elements of a system cannot be tested outright, simulation or digital twins are used instead. XR technologies offer great benefits when it comes to V&V as they can provide more detailed 3D models, diagrams, and other visual representations of systems which can help with system operation verification and validation.

Moreover, XR technologies enable the development of digital continuity across product life cycle management (PLM), enabling better management of hardware changes during operational life cycle stages. With XR technology, engineers can access virtual information repositories from anywhere, greatly speeding up the V&V progress.

Let's clarify the meanings of verification and validation once again. Verification involves checking whether a system or product has been designed and implemented correctly, ensuring that it functions as intended. It primarily focuses on confirming that the system meets specified requirements and adheres to design specifications. On the other hand, validation aims to determine if the system fulfills its intended purpose within its specific operational environment. Validation often involves assessing user expectations, usability, and overall effectiveness in meeting business objectives. Verification ensures that the system is built correctly according to defined specifications, while validation focuses on validating whether it meets users' needs and aligns with business objectives.

Virtual or digital mock-ups are a valuable tool that enables us to better understand a product and perform virtual testing of its functionality. By creating a digital twin of the system's behavior, we can parameterize it and accurately predict how it will behave in various scenarios. These virtual mock-ups allow us to explore different aspects of the product without having to build prototypes physically. This saves time and resources and allows for more efficient testing and analysis. By simulating the product's behavior in a virtual environment, we can identify potential issues or areas for improvement early on in the design process. This helps streamline development and reduce costs associated with physical prototypes.

Furthermore, digital mock-ups offer the advantage of being easily modified or updated as needed. Changes can be made quickly, allowing for iterative design improvements without significant disruptions to the production timeline. In summary, virtual or digital mock-ups provide designers and engineers with a powerful tool to visualize and test products before they are physically built. They allow for better awareness of product performance and enable accurate predictions of system behavior through parameterization.

How do we go about doing it? To achieve a successful synergy between experts involved in simulation and industrial metaverse in your organization, several steps can be taken:

1. Identify the key experts: Determine the individuals who are knowledgeable about simulation techniques as well as those familiar with the industrial metaverse. These experts will play a crucial role in ensuring that the V&V (verification and validation) activity meets your organization's requirements.

2. Establish effective communication channels: Set up regular meetings or communication channels where both groups of experts can collaborate, share insights, and exchange feedback. This will help bridge any gaps between their respective areas of expertise.

3. Define objectives and requirements: Clearly articulate what you aim to achieve through the V&V activity. This includes specifying what level of realism or abstraction is desired for the visual environment within the simulation. By aligning these objectives with input from both sets of experts, you can ensure that everyone is on the same page.

4. Promote knowledge sharing: Encourage knowledge sharing between simulation experts and industrial metaverse specialists by organizing workshops, training sessions, or cross-functional team projects. This will enhance their understanding of each other's domains and foster collaboration.

5. Regularly assess progress: Continuously evaluate how well collaboration efforts are working by monitoring project milestones, reviewing feedback from both teams, and assessing overall performance against established goals.

By following these steps to create a good synergy between simulation and industrial metaverse experts within your organization, you can ensure that the degree of realism (or abstraction) or the visual environment is suitable for the V&V activity.

XR Configurators

Figure 4-6. Illustration of collaborative design review with Immersive technology

Configuration management is a key capability to ensure the reliability and efficacy of industrial systems. The use of digital twins and 3D models can further improve this productivity by enabling the configuration of many features before the actual installation process begins. As shown in Figure 4-6, the configuration control is crucial in ensuring that changes made to a system or product are properly managed, documented, and approved to maintain consistency, reliability, and quality. Without configuration control, updates or modifications could result in unexpected impacts to the system, leading to unintended errors, downtime, or even safety risks. Configuration control also enables effective tracking and auditing of changes, allowing stakeholders to understand the current state of the system and maintain compliance with regulations and standards. Overall, configuration control helps mitigate risks, improve quality, and increase efficiency in managing complex systems and products.

Configurators provide companies in various industries with the ability to quickly assess how certain components should be implemented, what materials are needed, and how they should all fit together within a certain context. Particularly in the automotive industry, configurators allow for incorporating system-level configuration parameters against pre-defined requirements, ensuring that last-minute adjustments will not upset production lines.

Overall, configuration management is an invaluable factor in many industrial processes – especially when enhanced by digital twins – to help meet historical data accuracy requirements as well as being up-to-date with the latest technologies of XR-based models.

XR has already found applications in industries such as automotive where users are able to visualize the potential configurations of their vehicle before deciding to purchase it. This allows for improved customer experience while also simplifying the process related to configuration management. The interactivity offered by XR technology is far superior compared to traditional 2D configurators and is aimed at making the entire process of creating configuration items faster, more accurate, and efficient for any organization.

CHAPTER 4 CONVERGENCE

What happens if we lose control over our configuration? With or without the support of immersive technology, properly managing the configuration of technology systems is vital for the success and security of any business. Failing to do so can lead to disastrous consequences as we could be, in essence, breaking the digital continuity and losing control over our product specifications.

Note The Industrial metaverse holds the potential to enhance configuration management by introducing a third dimension. While having visually appealing 3D representations of customizable digital twins of products may seem enticing, it's important to focus on a key aspect: centralizing the configuration management repository. By centralizing the repository, it becomes harmonized and accessible across all phases of the product's life cycle and by different functions within an organization. This ensures that all relevant stakeholders have access to accurate and up-to-date information, enabling effective collaboration and streamlined processes. Having a centralized configuration management repository in the Industrial metaverse offers numerous benefits. It allows for better version control, reduces errors caused by outdated or conflicting information, facilitates traceability throughout the life cycle, and enables efficient decision-making based on real-time data. Therefore, while visual representation is valuable in enhancing understanding and communication, prioritizing a centralized configuration management system should be at the forefront when leveraging the capabilities of the Industrial metaverse.

CHAPTER 4 CONVERGENCE

Group Gamma – UML Diagrams and Functional Architecture

This group of use cases is a bit more speculative, as it aims to prove that XR can bring, yet unproven, advantages to functional analysis. UML Diagrams (Unified Modeling Language) are powerful tools for visualizing system architecture. They can be used by software developers and engineers to model the behavior of complex systems in order to design and review their own or peer-reviewed designs. In this section, we will be looking at the potential advantages that XR can bring to functional analysis in terms of UML diagrams (Figure 4-7). By combining 3D visualizations with interactivity, XR can open up new possibilities considering how architects and engineers approach system designs, both qualitatively and quantitatively. We will explore a range of use cases of how XR technology could help improve diagram accuracy without sacrificing review times or increasing development cost.

Figure 4-7. Abstract illustration of functional diagram in a 3D environment

CHAPTER 4   CONVERGENCE

The development of a methodology to integrate VR environments with SysML has shown promising results, but there are still a few areas where further extensions can be explored. One notable limitation is the current implementation of communication between SysML diagram visualizations and VR environments, which is currently one-way.

In the existing system, users can only update the SysML model and then export it for interactive visualization in the VR environment. This one-way communication restricts the ability of users to directly manipulate and interact with the diagram elements within the VR environment.

To enhance the functionality and architecture of this integration, future research could focus on enabling bidirectional communication between SysML diagrams and VR environments. This would allow users to not only update the model from within the VR environment but also manipulate and modify diagram elements directly in real time.

Such an extension would greatly enhance Model-Based Systems Engineering (MBSE) practices by providing a more immersive and interactive experience for system architects and designers. It would enable them to seamlessly transition between traditional diagram-based modeling in SysML and virtual simulations in VR environments, promoting better collaboration, understanding, and decision-making throughout the system development life cycle.

Virtual SysML

SystML is a specialized language derived from UML (Unified Modeling Language) that is widely used by systems engineers to model and analyze complex systems. It provides a comprehensive set of notations, diagrams, and semantics specifically tailored for system engineering tasks. By using SystML, engineers can visually represent the various components, interactions, and behaviors of a system. This allows them to better understand the system's structure and functionality, identify potential issues or bottlenecks early in the development process, and communicate effectively with stakeholders. SystML extends UML by incorporating additional elements that address specific system engineering concerns such as requirements management, parametric analysis, performance modeling, reliability assessment, and more. These extensions enable engineers to capture critical aspects of the system design within a single modeling framework. The use of SystML promotes consistency and standardization in systems engineering practices across different industries. It helps streamline the development process by providing a common language for different stakeholders to collaborate effectively on complex projects.

Based on the research conducted by different research groups, it appears that there is potential to develop a three-dimensional version of sysML. This immersive version could enhance systems engineering by introducing the third dimension to diagrammatic representations. This innovative approach may offer new possibilities and benefits for engineers working with SysML. In addition, it will allow the fruition of the functional architecture of the system in different details of complexity into the industrial metaverse.

Group Delta – Training and Field XR

An interesting, perhaps a bit obvious, benefit of XR to MBSE is the opportunity to develop training solutions that make full use of enhanced visualization to train the new generation of MBSE Engineers on the SE methods in an environment that can truly immerse the student in the digital continuity.

With the development of virtual reality, we have seen increased experimentation in the area of XR training and what roles it can play in workplace safety. Although the exact benefits of XR training are yet to be debated and proven with extensive use, it is already being embraced as a more enjoyable and effective way to conduct safety training for employees. Accessibility, convenience, cost-effectiveness, and flexibility – these all

CHAPTER 4   CONVERGENCE

are some of the potential advantages XR-based safety training has over traditional types. Additionally, employers also benefit from improved employee retention rates and employee engagement due to making use of such immersive learning technology.

For example, a company could use XR to provide virtual tours of its facilities, allowing employees to explore and learn about the company's operations without ever leaving the office, allowing customers to explore the space in a realistic way, to interact with the environment such as opening doors, turning on lights, and more. This topic is covered more thoroughly in Chapter 5, where different use cases are explored in detail.

Example of Field XR (Remote Assistance)

Virtual reality simulations are becoming increasingly popular for verifying systems and providing remote assistance. With technology making processes better and more efficient, virtual reality training is becoming an important aspect of routine operations. The key aspect of the matter is that the same simulations used for verification of the system can also be used for remote assistance, which helps to reduce costs associated with on-site visits or manual writing. Through this, employees are able to access information from any location, enabling them to achieve greater efficiency in their day-to-day tasks. This also allows companies to make use of standardized and augmented training manuals and documentation which further simplifies the process.

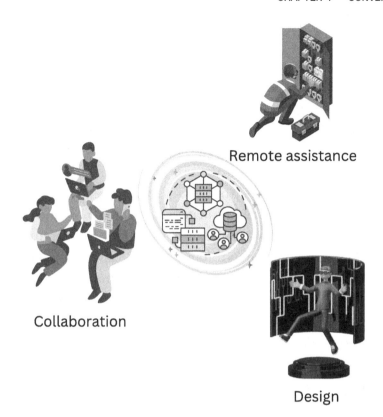

Figure 4-8. *Illustration of convergence of different XR activities*

For instance, XR can greatly assist with the maintenance and monitoring of cables underground. Technicians can quickly and accurately identify the location of cables without having to dig up the ground. XR technology can be used to create detailed 3D maps of underground cable networks, allowing technicians to quickly and accurately identify the location of any cables that may need to be replaced or repaired (Figure 4-8). This is especially useful in areas where cables are buried deep underground, as it eliminates the need for costly and time-consuming excavation.

CHAPTER 4 CONVERGENCE

Collaborative XR

In the last paragraph, we mentioned that by overlaying relevant maintenance manuals onto the physical environment, AR intuitively and vividly guides each step of on-site work. This cutting-edge technology provides a wealth of real-time information to maintenance personnel, simplifying their tasks and significantly improving work efficiency. With AR, maintenance personnel can access comprehensive instructions and visual aids directly in their field of view. They no longer have to rely solely on traditional manuals or training materials, which can be time-consuming to navigate. Instead, they can instantly access step-by-step guides and interactive elements through AR applications or wearable devices. The ability of AR to contextualize information in real time greatly reduces human error during equipment maintenance. Complex procedures are broken down into easily understandable instructions that are displayed within the technician's immediate surroundings, allowing them to focus on performing tasks accurately without constantly referring back to printed manuals. Moreover, AR enables remote collaboration between experts and on-site technicians. Through live video feeds or shared screens, experts can virtually guide technicians from anywhere in the world. This not only saves time but also allows for faster problem-solving when unexpected issues arise during maintenance operations. The immersive nature of AR empowers technicians with a deeper understanding of each task while fostering a safer working environment. This topic is also covered more extensively in Chapter 5.

A Proposed Metaverse MBSE Framework

We stated that the XR and MBSE provide an inclusive perspective by allowing for the integration of diverse stakeholders in the entire system life cycle. This perspective supports collaborative decision-making and iterative modeling, facilitating the incorporation of feedback from

CHAPTER 4   CONVERGENCE

individuals with varying needs and constraints. This encourages the creation of solutions that are accessible, affordable, and effective for all users. Overall, the XR MBSE framework helps to minimize the potential for exclusion and ensure that technology is broadly available and useful.

Finally, this inclusive perspective allows for the development of products and services that are tailored to the needs of all stakeholders, creating a more efficient and effective design, development, and deployment process.

While the possibilities of virtual reality seem endless, its implementation must factor in norms and regulations to ensure a safe and healthy product for industrial use. To further bridge this gap, we propose a framework for XR development to be used in industry that takes into account the specific tailorings required for successful industrialization.

This proposal aims to outline the necessary steps and considerations needed for successful implementation of an XR framework in industry, from navigating norms and regulations to the tailored aspects of integration with existing production lines. Such a framework will enable industries to fully harness the potential of extended reality and the metaverse while adhering to crucial norms and regulations.

The importance of simulations has been already emphasized. Simulations play a crucial role in various use cases discussed in this chapter. They provide valuable insights and help in training, testing, and optimizing processes within an industrial metaverse (Figure 4-9). To support an industrial metaverse, certain infrastructure and hardware are necessary. This includes systems capable of running complex simulations with high computational power. Additionally, specialized equipment may be needed to accurately replicate real-world conditions. Software development is a significant aspect of building a company-wide industrial metaverse. It involves creating tailored software solutions that enable simulations to be carried out effectively. This requires expertise in programming languages, algorithms, data analysis techniques, and user interface design. By leveraging simulations and investing in the

CHAPTER 4   CONVERGENCE

necessary infrastructure and software development efforts, companies can unlock the potential of an industrial metaverse. This can lead to improved efficiency, reduced costs, enhanced product quality, and accelerated innovation within various industries.

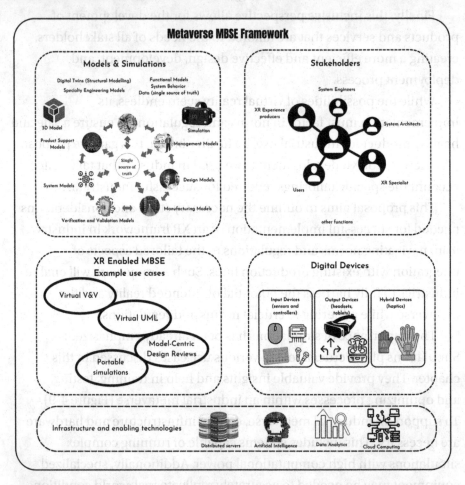

Figure 4-9. Key elements to build a Metaverse framework for your organization

114

CHAPTER 4   CONVERGENCE

A Brief Note on MBSE4XR

In this last section, the topic of MBSE for XR is covered, although less extensively. It is a given that SE is an enabling methodology for industrial processes. Especially in a market where users' expectations change so frequently like the XR.

For example, in the reference paper "Engineering Design Process for virtual reality headset" (see Sources), the principles of design of an XR headset are presented for a typical implementation. The reader is invited to read the relevant paper to have the full picture of the SE techniques employed.

The development of advanced user interfaces based on extended reality (XR) technology has enabled designers to accurately construct models from the ground up. This is a critical step in understanding the design, construction, and maintenance of XR infrastructure. MBSE also allows us to consider various scenarios and simulate its effects on societal acceptance. With new developments made in advanced visualization, users can even interact with complex simulations while finally designing real-world products. All these features come together when harnessing the power of MBSE techniques for design and operation, permitting designers to achieve much more accurate results with a great level of precision, and for the first time ever, with considerable reduced costs.

Therefore, it is important to design research that focuses on finding the best ways to create immersive experiences that are tailored to specific user needs, with interfaces and interactions that are easy to learn and use.

The Metaverse As System of Interest

Creating the cyberphysical infrastructure of the metaverse from scratch can indeed be a substantial undertaking. It would require significant investments and may pose challenges, particularly for small or medium enterprises. Developing a company metaverse necessitates not only specialized knowledge but also the ability to bridge the gaps between various databases. To embark on this endeavor, you would need to assemble a team with expertise in areas such as virtual reality, augmented reality, software development, database integration, and network infrastructure. This team could help you architect and build the necessary infrastructure for your metaverse. It's important to note that developing a custom solution can be time-consuming and costly. You would need to consider factors like hardware requirements, scalability (see interview with Cedric Ching in Chapter 7), data privacy and security measures, user experience design, ongoing maintenance, and more. Alternatively, if creating an entire metaverse from scratch seems daunting or unfeasible for your organization at present due to resource constraints or other limitations, you might explore leveraging existing solutions or collaborating with technology partners who specialize in building metaverses. This approach could save time and resources while still allowing you to achieve your objectives within a reasonable time frame. Ultimately, it is crucial to carefully evaluate the costs and benefits associated with each approach before making any decisions.

XR Devices As System of Interest

If your objective is to create XR devices, it is important to understand that the market for XR technology is constantly evolving and the needs of users are continuously changing. This makes it challenging to develop devices that can achieve massive adoption in industries. However, it's worth noting that successful XR devices often build upon existing business cases that have already proven to be financially successful. For example, devices focused on gaming, visual augmentation systems, or navigation systems have seen significant financial success. In this context, MBSE can be a valuable tool. By using modeling techniques and methodologies, MBSE allows for enhanced system understanding and documentation. This can help you in designing and developing XR devices by providing a structured approach to capture requirements, perform analysis, and make informed decisions throughout the development process. By leveraging MBSE principles alongside careful market research and a deep understanding of user needs, you can increase your chances of creating successful XR devices with broader adoption in the industry.

XR Applications As System of Interest

Systems engineering is a discipline that originated with software development and has since been applied to various industries, including development for XR applications. When it comes to immersive experiences, systems engineering can help in several ways. Firstly, it can assist in defining the requirements for the software by thoroughly understanding the needs and expectations of users. This includes identifying the desired level of immersion, interaction capabilities, graphics quality, and performance requirements. Next, systems engineering can aid in designing the architecture of the immersive software. This involves determining how different components or modules will interact with each other to create a seamless experience for users. It also includes considering factors such as compatibility with different platforms or devices. During development, systems engineering techniques like prototyping and iterative testing can be used to ensure that all aspects of the immersive experience are functioning as intended. This allows for early detection and resolution of any issues before they become significant problems. Managing complexity is another key aspect addressed by systems engineering. Immersive experiences often involve multiple layers of technology integration – from hardware devices like virtual reality headsets to sophisticated algorithms that generate realistic visuals and interactions. Systems engineering helps in coordinating these complex elements effectively. Moreover, as immersive software evolves over time through updates or new releases, systems engineering provides a framework for managing changes while maintaining high levels of quality and user satisfaction. In summary, applying systems engineering principles to immersive experiences as software allows for accurate life cycle mapping and effective management of complexity throughout the development process. It ensures that user requirements are met while maintaining high standards of functionality and performance.

Takeouts

- The Industrial Metaverse is a rapidly evolving field that uses extended reality (XR) to enhance system engineering practices. XR4MBSE and MBSE4XR are two terms used to describe the binomial convergence of two disciplines, AI4SE and SE4AI. XR4MBSE refers to extended reality applied to model-based systems engineering practices, utilizing virtual simulations and augmented reality to aid engineers in verification and testing at multiple stages of the process. XR is becoming increasingly popular for its ability to improve accuracy, efficiency, and cost savings for engineering operations. MBSE4XR is the use of MBSE to enhance XR technology, involving all modeling techniques used to develop XR systems. Implementing an XR framework requires careful consideration of factors such as research and analysis, stakeholder engagement, risk assessment, customization, training programs, user experience testing, compliance monitoring, data security measures, scaling up possibilities, and industry collaboration.

- Digital continuity and life cycle mindset are crucial for creating a metaverse, enabling smart connected products, new business models, financial transparency, and faster time to market. Key enablers include infrastructure, data distribution systems, device interfaces, and XR applications. Infrastructure ensures seamless data transmission and processing, while interfaces allow for interoperability and seamless user experience. Product Lifecycle Management (PLM) is a systematic approach to managing product changes from design to disposal. PLM software applications automate data management and integrate it with other business processes. Emerging disciplines like Digital Engineering, Model-Based Engineering, Model-Based Systems Engineering, and Architecture are essential for incorporating immersive technology into organizations. MBSE Virtual Engineering is a methodology that is divided among professionals, but a reasonable extrapolation from current figures can help create a suitable team to tackle challenges in XR-based virtual engineering.

- A systems architect is responsible for designing and developing computer networks and systems for company operations, conducting research, devising strategies, improving existing systems, and implementing solutions for optimal processes. They must monitor implementation progress, provide technical support, and train new workforce members. An MBSE engineer with XR expertise works in IT Agile/Waterfall environments, responsible for technical analysis, planning, architecture, and design of solutions. An XR producer manages programmers, developers, and day-to-day processes, while a virtual or digital engineer provides a user-centric perspective. An MBSE-XR champion is an early adopter of XR technology, providing feedback to teams in charge of improving immersive experiences. Narrative designers play a crucial role in creating immersive experiences within the Industrial Metaverse, combining storytelling, design, engineering, simulations, and model-based systems.

- The current model-based approach to industrial system design has limitations, such as a lack of continuity in Product Lifecycle Management (PLM). To address these issues, deep modeling in a virtual environment can be used to simulate various activities, leading to more efficient and effective product development. XR can bring numerous benefits to a model-based system thinking approach and the life cycle, including immersive mock-ups and digital twins. Digital twinning, augmented reality, and virtual reality can be used to create realistic models for development, testing, and validation. XR can also enhance industrial processes, such as zero defect manufacturing and automatic simulations. However, these methods do not fully utilize advanced data-driven technologies, which are essential for Industry 4.0.

- Zero Defect Manufacturing (ZDM) is a modern approach that uses digital technologies like Artificial Intelligence (AI), Machine Learning (ML), and large artificial datasets to anticipate and prevent issues in product and process scenarios. This enhances autonomy and quality control, incorporating predictive outcomes and facilitating comprehensive feedforward and feedback control loops. The introduction of Extended Reality (XR) technology can enhance process performance by simulating processes and providing continuous feedback. XR applications can improve assembly processes, logistics planning, and logistics support, while automatic simulations allow organizations to create immersive experiences and make informed decisions. Collaborative Design Reviews (CDRs) are also using XR to enhance collaboration and save time. Overall, incorporating XR into PLM workflows can streamline development processes and improve product quality.

- Immersive technology in design reviews requires digital continuity, which includes data, simulations, and models. XR simulations are used for verification and validation (V&V) in systems engineering, providing detailed 3D models and visual representations. They enable digital continuity across product life cycle management (PLM), allowing better management of hardware changes during operational life cycle stages. Virtual or digital mock-ups provide a deeper understanding of a product, allowing for efficient testing and analysis. To achieve successful synergy between simulation and industrial metaverse experts, identify key experts, establish effective communication channels, define objectives and requirements, promote knowledge sharing, and regularly assess progress. This ensures that the degree of realism or visual environment is suitable for the V&V activity.

CHAPTER 4   CONVERGENCE

- Configuration management is crucial for industrial systems' reliability and efficacy, and digital twins and 3D models can enhance this by enabling feature configuration before installation. Configuration control ensures changes are properly managed, documented, and approved, maintaining consistency, reliability, and quality. Configurators, particularly in the automotive industry, allow for quick assessment of components and materials, ensuring compliance with regulations and standards. XR technology can improve configuration management by allowing users to visualize potential configurations before purchasing, simplifying the process. The Industrial metaverse can enhance configuration management by centralizing configuration management repositories, ensuring accurate information, facilitating collaboration and streamlined processes. XR can also improve UML diagrams in functional analysis by combining 3D visualizations with interactivity, enhancing Model-Based Systems Engineering practices. Future research could focus on enabling bidirectional communication between SysML diagrams and VR environments, promoting better collaboration and decision-making throughout the system development life cycle.

Questions

- Can you explain the concepts of XR4MBSE and MBSE4XR in the context of system engineering practices?
- What are some examples of new business models enabled by digital continuity and life cycle mindset?

CHAPTER 4   CONVERGENCE

- What is the role of an MBSE engineer with XR expertise in IT Agile/Waterfall environments?

- How can deep modeling in a virtual environment improve Product Lifecycle Management (PLM)?

- What is immersive technology in design reviews and how does it relate to digital continuity?

- How do digital twins and 3D models enhance configuration management in industrial systems?

CHAPTER 5

Key Use Cases for XR

Companies that successfully implement XR have a clear business problem they are looking for or need to solve. This statement seems obvious but it's surprising how often companies and individuals (even the experienced innovation managers and teams found in large companies) become excited by new hardware and software and will happily purchase to test and play.

The chapter explores the use of Extended Reality (XR) technologies like Virtual Reality (VR) and Augmented Reality (AR) in business. It emphasizes the importance of identifying specific business problems before implementing XR solutions. Combining different XR use cases can create unique solutions tailored to an organization. VR can enhance work processes, facilitate prototyping, training, and performance monitoring, leading to cost savings and improved productivity. AR-based remote assistance became crucial during the COVID-19 pandemic, allowing experts to provide real-time guidance to on-site workers. Immersive training with XR can enhance retention and practical skills. XR guidance provides step-by-step instructions, reducing errors and improving productivity. AR enables real-time, 3D collaboration on digital assets. XR technologies can revolutionize maintenance by providing tools for diagnostics, training, and realistic simulations, enhancing safety and efficiency.

CHAPTER 5   KEY USE CASES FOR XR

VR Applications in Your Work

Companies that successfully implement XR have a clear business problem they are looking for or need to solve. This statement seems obvious but it's surprising how often companies and individuals (even the experienced innovation managers and teams found in large companies) become excited by new hardware and software and will happily purchase to test and play.

It's critical that careful consideration and analysis of problems and potential solutions take place before purchasing XR solutions. To help understand the types of problems that can be solved, there are a number of well-established, industry understood, use cases that have been delivered by various companies across the globe.

A word of warning, real-world business problems do not fit into clear-cut use cases. In reality, it's likely that various use cases elements will need to be combined to solve a solution that's unique to your organization. However, having an understanding of the opportunities and problems that can be solved, helps companies to focus their projects, create clear features and requirements and ensure that the implementation of XR is successful. Solving business problems which provide tangible benefits and return on investment is crucial and will go a long way to ensuring future funding and support.

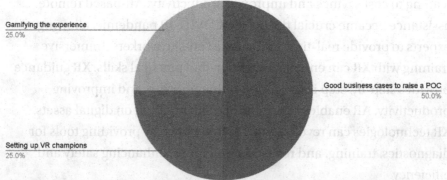

Figure 5-1. Case study. VR applications in your work (POC stands for Proof of Concept)

CHAPTER 5   KEY USE CASES FOR XR

AiShed conducted a survey among a pool of aerospace engineers (Figure 5-1). It shows that introducing virtual reality (VR) into the workplace can bring numerous benefits and enhance various aspects of work processes. To convince your colleagues to embrace VR, it is essential to highlight the compelling business cases that demonstrate its value.

One effective approach is to propose setting up VR champions within your organization. These individuals can act as advocates for VR technology, showcasing its potential applications and benefits across different departments. By having dedicated VR champions, you can create awareness and generate interest among your colleagues.

- Another persuasive argument is the ability of VR to facilitate prototyping, refinement, and industrialization processes. With VR, teams can create virtual prototypes of products or designs, allowing for quick iterations and improvements before investing in physical prototypes. This not only saves time but also reduces costs associated with traditional prototyping methods.

- Furthermore, VR offers a gamified experience that can make training sessions more engaging and effective. By incorporating interactive elements in virtual environments, employees can acquire new skills or undergo simulations in a more immersive manner. This enhances learning outcomes while providing a safe space for practice without real-world consequences.

- Additionally, monitoring performance through XR (extended reality) technologies such as augmented reality (AR) or mixed reality (MR) can offer valuable insights into employee productivity and efficiency. By visualizing data overlays or providing real-time guidance through AR/MR headsets, workers can optimize their workflows and make informed decisions based on contextual information.

CHAPTER 5   KEY USE CASES FOR XR

Ultimately, by presenting these business cases highlighting the advantages of using VR in various work scenarios – from training to prototyping and monitoring – you can raise a proof of concept (POC) that demonstrates its potential impact on productivity, cost savings, employee engagement, and innovation within your organization (Figure 5-2).

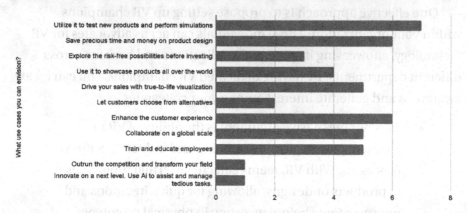

Figure 5-2. *Case study. Use cases you can envision*

The use of XR (Extended Reality) technology offers a wide range of possibilities across various industries. This second chart highlights the numerous use cases where XR technology can be applied, providing businesses with significant advantages.

One notable application is simulations, as said in Chapter 2, where XR technology allows companies to create virtual environments that replicate real-world scenarios. This enables organizations to save money by conducting risk-free simulations before implementing costly projects or processes.

XR technology also serves as a powerful tool for showcasing products and services. Through immersive visualizations, businesses can present their offerings in a more engaging and interactive manner, enhancing customer experiences and driving sales. Lastly, embracing XR technology provides businesses with a competitive edge in today's fast-paced market. By incorporating immersive experiences into their operations, companies can differentiate themselves from competitors and captivate customers with unique and memorable interactions.

Overall, this chart showcases the vast potential of XR technology across multiple industries by enabling simulations, saving costs, offering risk-free showcases, enhancing visualization for training purposes, facilitating collaboration among teams, and providing a competitive advantage in the market.

Remote Assistance

Due to the COVID pandemic, Remote Assistance has become the most well-known and probably widest implemented use case. Many companies faced the challenge of fewer people being allowed into the factory or place of work, but critical business activities had to continue. Certain industries needed to continue to manufacture and maintain services, which often required staff's physical presence. If a problem occurred (e.g., a breakdown, change of configuration, or part, etc.) a skilled worker (someone who had knowledge and experience to solve the problem/make the change) may not be in attendance. This may lead to downtime which increases costs and reduces efficiencies.

The Remote Assistance use case involves having remote subject matter experts assisting on-site workers by using AR. When a problem or situation for which the documentation is not satisfactory, or a more detailed inspection is required, expertise may be required from someone who is not on site. By using Remote Assistance, any worker can connect with an expert (who can be anywhere in the world) who will be able to support. This can also solve the critical aspect of "time of intervention," as it is not always possible to have the expert travel to the needed location on time for the issue to be sorted without disrupting the workflow. It is quite obvious that having an expert on site has many advantages, an expert can provide valuable insight into what is needed in order to achieve a successful outcome. In the past, this was achieved by traveling to remote sites, video conferences, and other telecommunication means.

CHAPTER 5   KEY USE CASES FOR XR

With the use of AR, real-time instructions, information and questions can be provided by the remote expert which will guide the worker without requiring the expert to be present in person (Figure 5-3). The expert can annotate and enhance the worker's view by overlay's including images, animations and/or videos. Without this support, the problem may not be completed (on time or according to guidelines).

Figure 5-3. *Use of XR devices in an industrial context*

Training

Introducing immersive training in an organization is indeed a process that should be approached gradually and thoughtfully. It is essential to consider the diverse demographics within the workforce, as different individuals may have varying levels of comfort and familiarity with technology. Convincing people to invest significant time in an unfamiliar environment can be a challenge, but there are several persuasive factors to consider.

CHAPTER 5   KEY USE CASES FOR XR

- Firstly, emphasize the benefits that immersive training can bring to both individual employees and the organization as a whole. Immersive training allows for hands-on experience in realistic scenarios, providing valuable practical skills and knowledge. This kind of learning has been proven to enhance retention rates and improve overall performance.

- Secondly, address any concerns or reservations about technology by highlighting its user-friendly nature.

- Assure employees that immersive training programs are designed with simplicity in mind, making them accessible even for those less technologically inclined.

- Offer support resources such as tutorials for introductory sessions to help ease any technological anxieties.

- Additionally, consider tailoring the introduction of immersive training based on employee preferences and learning styles.

- Provide options for blended learning approaches where traditional methods can coexist with immersive experiences. This way, individuals who may be hesitant about fully embracing virtual environments can still benefit from supplementary materials or alternative methods. It is also crucial to involve employees throughout the process by seeking their input and feedback actively.

- Encourage open dialogue so that concerns can be addressed promptly while showcasing how their involvement will contribute directly to shaping future training initiatives.

CHAPTER 5　KEY USE CASES FOR XR

- Lastly, demonstrating success stories from other organizations that have implemented immersive training can greatly influence skeptical employees' perceptions. Sharing real-life examples of improved outcomes resulting from this type of training helps build confidence in its effectiveness.

By taking these persuasive steps into account (Figure 5-4) – emphasizing benefits, addressing concerns through support resources and tailored approaches – organizations can gradually overcome resistance toward new technologies and foster a positive acceptance of immersive training among all employees.

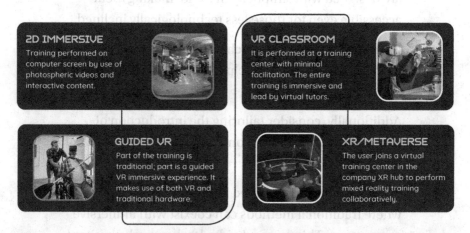

Figure 5-4. A view of the possible gradual implementation of immersive training in a professional organization

Note: As tolerance levels toward this form of learning can vary among individuals, it becomes crucial to consider different perspectives when discussing VR education. The topic of cybersickness is covered in Chapter 6 when deployment is also covered.

As said in Chapter 4, XR-enabled training can be delivered to groups or individuals. The training material and assets can be delivered via various hardware platforms (e.g., tablet, XR wearable, PC) and can reuse learning assets stored in a corporate knowledge base. One of the key benefits of XR training is the trainee's interactions, skill development, and achievement of competency can be captured in real time (retention rate). The recording or data about the competency level can be stored in the learning management system.

XR Training is delivered with real machines or products, the trainee can see and hear the real world circumstances and interact with equipment or other people directly, increasing the kinesthetic learning opportunity.

A head-worn, XR-enabled training solution can create highly compelling and interactive learning materials for playback as video or by other XR-enabled display users. The trainer can use gestures and enhance the training materials with text, graphics, animations, or videos.

In the future, a key advantage of XR training is the potential to assess the impact of the training by using feedback from the other trainees, remote instructors and other stakeholders. This has been already addressed in Chapter 4. More insights can be also found in the interviews presented in Chapter 7.

Survey Conducted by AiShed

In collaboration with AIShed, a survey was conducted to evaluate the attitude of engineers toward XR training. The results of this survey shed light on the current perception and acceptance of XR training among engineers. By understanding their attitudes toward this technology, it becomes possible to identify any barriers or challenges that may hinder its widespread implementation in industry. These findings serve as a foundation for further research and development efforts aimed at maximizing the potential benefits of XR training in industry.

Furthermore, these insights also help establish the requirements for the successful implementation of XR training in various engineering sectors. By considering factors such as hardware compatibility, software integration, user experience, and content development, organizations can effectively leverage XR technology to enhance training programs and improve overall productivity.

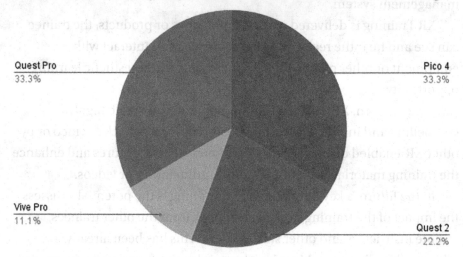

Figure 5-5. Case study. What is the type of device to be chosen for XR training?

From this first chart (Figure 5-5), we can conclude that there is no clear preference for the device to be utilized, as long as it has the necessary power to run the training environment without lag. Ergonomics is also important but rather subjective at this stage of development. There is a slight prevalence of Pico4 and Quest Pro since those are the most up-to-date devices available at the time of writing. This result is not surprising since at present no hardware is clearly dominant on the market, and R&D is ongoing.

CHAPTER 5 KEY USE CASES FOR XR

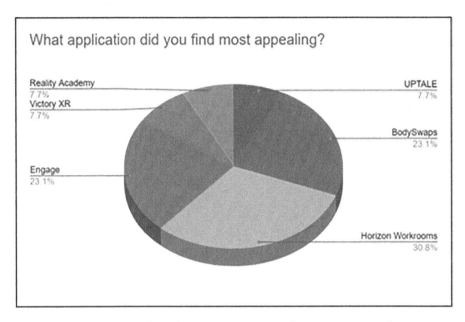

Figure 5-6. *Case study. Choose your virtual environment for training*

The second chart (Figure 5-6) highlights the preference for the Meta solution "Horizon Workrooms." This preference can be attributed to several factors. Firstly, Horizon Workrooms is likely the most mature solution among the options presented. This means that it has been developed and refined over a longer period of time compared to other solutions such as Engage and BodySwaps. The maturity of a solution often translates to a higher level of reliability and functionality.

Secondly, Horizon Workrooms may have a larger development team behind it. A larger team can contribute to faster innovation and more frequent updates or improvements to the solution. This can result in a more robust and feature-rich offering that appeals to users seeking advanced capabilities.

135

CHAPTER 5 KEY USE CASES FOR XR

It's important to note that while Horizon Workrooms may currently be favored based on these factors, preferences in this space can evolve rapidly as technology continues to advance. Therefore, it's essential for users to stay informed about new developments and emerging solutions that may better meet their specific needs.

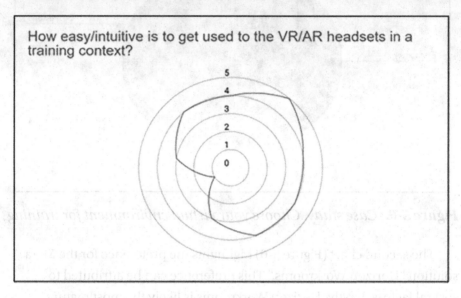

Figure 5-7. *Case study. The usage of devices. Intuitive or difficult?*

According to the chart in Figure 5-7, a majority of the interviewees reported finding the use of VR headsets relatively easy. This indicates a positive level of awareness and comfort with this technology among the participants. The ease of use of VR headsets is crucial in various stages of development and implementation. During the prototyping phase, when new VR applications are being tested, it is important for users to be able to easily navigate and interact with the virtual environment.

Similarly, as VR experiences go through refinement and industrialization processes, ensuring that users can easily adapt to using VR headsets becomes even more important. This allows for smoother integration into various industries such as gaming, education, healthcare, and more.

CHAPTER 5　KEY USE CASES FOR XR

Furthermore, ongoing monitoring of user experiences with VR headsets is essential for identifying any potential challenges or areas for improvement. By understanding how users perceive and interact with this technology, developers can make necessary adjustments to enhance usability and overall user satisfaction.

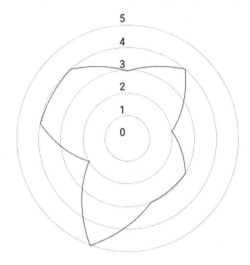

Figure 5-8. *Case study. Relevance of XR training*

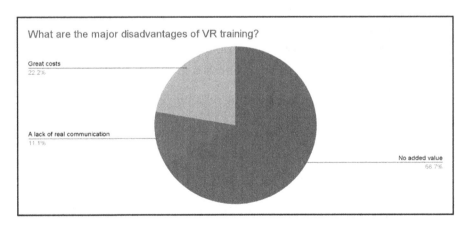

Figure 5-9. *Case study. Disadvantages of XR training*

The charts (Figures 5-8 and 5-9) presented provide valuable insights into the level of interest in XR training among the interviewees. It is evident that there is a low level of interest, which could be attributed to the fact that most of the interviewees were system engineers. The discipline of system engineering has traditionally been focused on documentation rather than hands-on training or immersive experiences like XR. This could explain why there is a lack of awareness and enthusiasm for XR training among this particular group.

However, it's important to note that this chart only represents a specific and limited sample of individuals and may not reflect the overall sentiment toward XR training. It would be beneficial to conduct further research and gather data from a more diverse range of professionals to obtain a comprehensive understanding. In order to increase interest and adoption of XR training, it may be necessary to emphasize its benefits in terms of prototyping, refinement, industrialization, and monitoring processes. Highlighting how XR can enhance these areas within system engineering could help generate more enthusiasm for incorporating XR technologies into training practices.

Additional remarks from the interviewees:

- "Headsets were comfortable, but the training apps were poorly thought out, and the controllers were very intermittent/laggy, meaning it often took ten attempts to select something. This is too frustrating a user experience to convince businesses to invest in VR."

- "It requires large scale testing on a bigger population of engineers."

Guidance

In the evolving landscape of workforce dynamics, companies find themselves grappling with transformative shifts. Traditional manufacturing enterprises, in particular, face challenges such as an aging workforce, an escalating demand for flexible work arrangements, and a struggle to attract younger talent.

This chapter explores a compelling XR Guidance use case that offers a solution to these challenges by delivering contextually-triggered XR-enhanced step-by-step instructions for a myriad of tasks.

In the absence of XR technology, disseminating accurate instructions to workers at precisely the right moment with clarity and precision poses a formidable challenge. Especially for workers who haven't memorized the steps, such as expert workers, the need to locate information often leads to increased downtime in task completion.

Tasks involving new tools or unfamiliar processes may require assistance in finding, calibrating, and operating machinery or tools. The presence of numerous physically similar objects, distinguishable only upon close inspection, amplifies the risk of human errors without a reliable system for automatic recognition and tracking of parts and tools. To mitigate errors, workers must meticulously inspect all components, a time-consuming process.

While workers may have undergone prior training (refer to the training use case), the incorporation of highly contextual guides, akin to having a trainer by their side, becomes essential. XR-enhanced guidance proves invaluable, especially when using new tools or executing unfamiliar processes, by confirming the correct completion of each step before guiding the worker to the next one.

XR guidance experiences can exist as stand-alone applications, running entirely on the worker's device (such as a tablet or wearable), or they can be seamlessly integrated with the company's standard operating procedures repository.

The primary advantages of XR-assisted guidance manifest in the substantial reduction of time and errors associated with task performance. Beyond these core benefits, improvements in employee productivity and satisfaction are notable outcomes. In specific instances, particularly when dealing with unique and infrequent instructions, XR-assisted guidance alleviates a worker's cognitive load, leading to enhanced performance and satisfaction.

Additional potential benefits encompass a decrease in training time and an overall increase in employee productivity, even in cases where specific task training has not been undertaken.

Assembly

In the contemporary landscape of intricate product designs and diverse manufacturing lines, companies are increasingly turning to XR to navigate workers through complex and infrequent tasks. Technicians often grapple with complex assembly procedures presented in traditional forms such as printed manuals or on-screen displays with illustrative content, including animations. These documents serve not only as guides but also as records of completed steps. Juggling the task at hand, the components involved, and the instructions source, whether in print or electronic form, demands the technician's constant attention.

While routine and straightforward tasks that lack human decision-making or dexterity requirements are often delegated to existing robotic systems, there exists a cost threshold for automation. Companies evaluate the balance between the cost of equipment and supplies vs. human labor, considering the potential advantage of assigning additional duties to technicians.

CHAPTER 5   KEY USE CASES FOR XR

As assembly complexity increases, the need for technicians to operate or supervise machines becomes more pronounced. The introduction of instructions depends on the frequency of technician-machine interactions and the associated duties. Machines may also integrate with factory control systems to monitor crucial usage details.

Certain intricate assembly processes may not be conducive to mechanized support due to prohibitive costs or the need for technician involvement in evaluating and verifying components or materials. In such cases, the assistance of a technician or alternative machinery is indispensable.

AR-enabled systems, coupled with display devices linked to assembly management systems and work order documentation, offer a transformative solution. Technicians, guided by AR assistance, receive real-time step-by-step assembly instructions directly in their workspace. The choice of AR display is influenced by factors such as the technician's need for hands-free operation, available space for additional screens, and compatibility with emerging display devices like wearable AR or projection AR.

With AR assistance for assembly, technicians no longer need to shift their focus between the task and documentation. Each step in the complex assembly process is automatically detected by the system or manually selected by the technician.

The benefits of AR-assisted complex assembly are quantifiable, translating into shorter procedure times with fewer errors. By eliminating the need to divert attention from the work area to documentation, technicians experience a reduced cognitive load. This streamlined approach proves advantageous even for technicians without specific training in individual assembly duties, leading to time savings on training and an overall boost in productivity.

We encourage you to find out more through the perspective presented by Dr Laughlin in Chapter 7.

CHAPTER 5   KEY USE CASES FOR XR

Collaboration

This use case focuses on AR support for two or more persons interacting with digital assets for collaborative design, creation, and review while nearby or at a distance. One or more of the following qualities are considered to be signs of collaboration:

- Two or more individuals can converse in real time, and everyone can view the same data at once. Participants in a session can also comment on, annotate, and modify shared digital files.

- The shared digital assets are kept in a location from which one or more users can obtain them both during the current session and in the future.

- Reaching consensus, co-creating new digital assets or material, or making decisions are just a few of the collaboration's objectives.

People work together in many different ways and with many different tools. People that are physically close to one another cooperate utilizing gestures, one or more physical models (such as scale replicas), electronic or physical boards, and writing implements. Platforms like WebEx, Google Meet, GoToMeeting, Aptero Teams add-on, or other similar commercial services can facilitate remote collaboration. In some scenarios, adding information or annotating shared assets requires the usage of a mouse and/or keyboard by the collaborators.

CHAPTER 5   KEY USE CASES FOR XR

> "I liked the 'screen' in the room with data on it, I had a vision where it might be good to run reviews in such a place but with more screens with presentations/calculations. We've used the room with many screens for this kind of thing, making this possible remotely might be interesting. Based on some initial discussions the challenge I think will be value 'I can do that with Meet' or 'I don't see the point vs. presentation' vs. the cost but I'll find out more as I talk to more people."

Collaborators employing common technology solutions are nearly always restricted to 2D graphics and text without support for augmented reality. During a session, a notion may occasionally be "brought to life" with the help of video and animation. As an alternative, there are places where people can work together and co-create utilizing 3D models while donning a Virtual Reality headset. Participants in these situations share a totally digital environment and its items.

Figure 5-10. Example of virtual collaboration (Source: Aptero)

CHAPTER 5 KEY USE CASES FOR XR

Collaborative AR use cases cover the gap left by current collaborative platforms when participants in a review, design, or problem-solving session must engage with both real-world objects and digital assets in real time and in three dimensions (Figure 5-10). Users might not be aware of the collaborators' relative positions in relation to the digital assets without AR.

By integrating physical and digital things into their respective contexts, participants in collaborative AR systems may be able to shorten or do away with the time and expense of travel to share a physical location.

AR for collaboration is a highly advanced use case when compared to other use cases because it involves not just advanced technologies but also users (facilitators) who are trained and experienced in manipulating digital objects in AR experiences.

Digital assets are imported into a "common" location before an AR session that involves collaboration. These use cases are especially advantageous for 3D models since they may be edited, changed, and annotated in real time. By observing where their collaborators stand in relation to the digital assets and themselves, participants can provide insights regarding problems or fresh suggestions. The advantages of augmented reality (AR)-assisted collaboration can largely be quantified in terms of the necessity for two or more users to travel to the same physical location in order to fulfill their objectives.

Additionally, there might be advantages because utilizing this platform is less cognitively demanding than using conventional collaborative platforms, and users are more satisfied overall. The movement of a digital asset while it is being viewed by numerous people online is comparable to the movement of a scale model when it is present in the same physical location as users.

Navigation

The professional is frequently required to work swiftly and safely in a location that they have never been to or are unfamiliar with. The navigation focuses on using augmented reality (AR)-enabled technologies that show symbols or provide audio cues that are synchronized with the outside environment to guide a person or individuals between destinations without delay or safety hazards. In the Industrial Metaverse, navigation is based on a 3D environment, allowing users to move around in a virtual space. This is in contrast to the Internet network, which is based on a 2D environment, where users are limited to scrolling up and down a page. By integrating avatars, virtual reality headsets, and augmented reality glasses, this immersive technology not only facilitates faster decision-making processes but also enhances the customization options. For example, the ultimate goal for an airline company is to ensure optimal passenger comfort and overall operational efficiency. Simulating the flight path, the aircraft's performance, and the weather conditions, an airline company can establish what is the highest level of passenger comfort and operational efficiency.

In general, navigation in the Industrial Metaverse is based on

- Immersive experience, with users able to interact with objects and other users in the virtual space. This is in contrast to the Internet network, which is based on a more passive experience, where users are limited to clicking on links and viewing web pages.

- Intuitive experience, with users able to use natural gestures and movements to interact with the virtual environment. This is in contrast to the Internet network, which is based on a more structured experience, where users are limited to typing in keywords and navigating through menus.

- Secure experience, with users able to access the virtual environment without the risk of malicious attacks. This is in contrast to the Internet network, which is based on an open platform, where users are vulnerable to malicious attacks.

The objective of virtual navigation is assisting a worker who is traveling on foot, with the use of a piece of industrial equipment or a vehicle, to go securely and efficiently between two sites is the definition of navigation. For instance, in factories or warehouses, a worker may need to avoid areas that may be physically separated from one another by walls, other obstacles, safe entrances, and other barriers. When a professional uses an AR-enhanced, network-connected system to receive a request to travel to a job site or destination, the route is created in real time by artificial intelligence using all of the most recent weather data available to the algorithms. The expert can then understand turn-by-turn the user arrives at the destination, the device also detects their position and is instantly turned off. Even when the destination is known, using previous trajectories as a guide when traveling there is not necessarily secure and dependable. Following a previous path is not always practicable due to hazardous or impassable conditions that a professional may have already navigated. On the other hand, a previous route might have accounted for circumstances that have subsequently changed, making a shorter, more effective route possible. The professional may need to stay away from areas under cranes where there are working machines or where building materials have just been dumped.

The kind of AR navigation display utilized relies on a variety of elements, including

- The necessity of using both hands, such as when driving a car
- The device's size and weight, as well as any additional items that must be worn or carried to the location

- Connectivity to the main navigational systems containing all live data about the user's environment

It is possible to connect an AR-enabled navigation system with other use cases so that the user receives turn-by-turn instructions either before or after completing a job. The professional in travel can pay more attention to acquiring other information or being aware of their surroundings by using AR navigation, which frees them up to do so. The system either automatically detects the initial and every consecutive step taken on the route established by the system, or the technician must manually dismiss each step. The sounds and/or symbols that indicate the route are displayed and registered at each step. When a professional arrives, the system instantly recognizes their status and can verify their arrival at the designated place in the management system. Alternatively, the technician can communicate with the AR system by voice, gesture, or some other means to confirm arrival.

Some job sites have clear boundaries, and the technician might be given maps. Additionally, in some highly complex environments, such as buildings under construction, even a digital map on an app that is not continuously updated may not be current and reflect all conditions. The user must search in the general area of landmarks for other locations that are either vaguely defined or unknown. In some use scenarios, as in a warehouse, the exact orientation and elevation (together with 2D position/location) are crucial to minimize user search time for a part or item or discovering a broken pipe in the field.

Therefore, the advantages of AR-enhanced navigation include giving the user real-time directions that do not conflict in any way with their perception of the outside world and allowing them to follow the route without having to look away from their surroundings. The route is designed to ensure the least amount of downtime while maintaining the highest level of safety. Real-time mapping of locations and facilities is another potential feature of a navigation system enhanced by augmented

CHAPTER 5   KEY USE CASES FOR XR

reality. For the benefit of other users, an updated 3D model of the real environment is created using the camera or other sensors on the AR display along with real-time positioning. Without giving any thought to the environment or their destination, the professional is adding to the cloud-based global map. Without Augmented Reality support, a user may be provided a printed document (e.g., a map) or an app on a smartphone. In either electronic or print formats, the technician must focus attention on both the real world and the map to follow turn-by-turn instructions.

Figure 5-11. Example of Augmented Reality navigation

Professionals are more productive overall when AR-enabled navigation is accessible (Figure 5-11), regardless of whether they have visited the destination before and regardless of the surrounding area (indoor, outdoor, underground, etc.), receiving instantaneously other types of information (not related to navigation) while in transit, since they, going to a location, will need to evaluate the safety of proposed routes before following them.

CHAPTER 5　KEY USE CASES FOR XR

Maintenance

The purpose of this use case is to check processes, products, and workplaces for safety and quality assessment or documentation used in the maintenance activities. It makes sure that all regulations and policies are followed and aids in the detection and/or reduction of a wide variety of different types of risks.

By using Industrial Metaverse technology for maintenance activities, companies can reduce the risk of accidents, improve safety, and increase efficiency (Figure 5-12). Such a technology is revolutionizing the way of maintenance and is providing maintenance professionals with powerful tools to help them do their jobs more efficiently and effectively. This use case focuses on using XR, AR, and VR systems to diagnose problems to perform troubleshooting and, if problems are found, to give the user digital tools and visual instructions for completing repair and maintenance chores.

Figure 5-12. *An example of AR visualization for maintenance*

149

CHAPTER 5   KEY USE CASES FOR XR

XR technology can be used to

- Create virtual simulations of a facility or equipment, allowing maintenance professionals to practice and troubleshoot in a safe, virtual environment
- Create interactive 3D models of a facility or equipment, allowing maintenance professionals to quickly identify and diagnose problems
- Create interactive visualizations of a facility or equipment, allowing maintenance professionals to quickly identify and diagnose problems
- Create interactive training modules for maintenance professionals, allowing them to learn and practice maintenance tasks in a safe, virtual environment
- Create interactive maintenance checklists, allowing maintenance professionals to quickly and accurately identify and address maintenance issues

AR can be used to

- Provide technicians with real-time data on the condition of equipment, such as temperature, pressure, and vibration. This data can be used to identify potential problems before they become serious.
- Provide technicians with a 3D view of the environment, allowing them to identify potential hazards and take corrective action.

VR can be used to

- Simulate the environment, allowing technicians to practice their inspection skills in a safe and controlled environment.

CHAPTER 5   KEY USE CASES FOR XR

- Provide technicians with a virtual tour of the facility, allowing them to identify potential safety issues and take corrective action.

Realistic Experiences

Such a use case can provide an overlap with other visualization use cases. The objective of the realistic experience is to create a real-world-like environment, analyzing the behavior of a system or to predict the outcome of a certain situation (Figure 5-13).

Figure 5-13. Metaverse positioning

In this use case, 3D models are used to replicate the insertion or relocation of objects within the real world and in interaction with it. The 3D models can represent weather patterns, energy flows, industrial equipment, infrastructure (like HVAC), or moving objects in a zone or confined space (like a factory). With the use of an XR device such as a tablet, users can now click an element by looking at it and pinching two fingers together, move the element by moving their pinched fingers, and scroll by flicking their wrist. This type of gesture-based control can eliminate the need for a physical keyboard or mouse, making it easier to use digital content on the go.

Gesture-based controls are becoming increasingly popular in a variety of applications and are sure to become even more commonplace in the future. They can also represent complex processes that require employees to lift heavy objects or carry out tasks with unusual shapes, as well as serious games (which overlap with training) and simulations of packing various objects into a volume (i.e., for shipping). Additionally, a realistic experience can be utilized in skill development (training) use cases, where the user can interact with the environment by using a controller or a headset.

NOTE: It shall be clear that a realistic experience is a computer-generated environment that allows users to interact with a virtual world, while all simulated behaviors of the analyzed system are represented by using a mathematical model.

Takeouts

Extended Reality (XR) Technologies: The document explores the use of XR technologies like Virtual Reality (VR) and Augmented Reality (AR) in business, emphasizing the importance of identifying specific business problems before implementing XR solutions.

Applications and Benefits: XR can enhance work processes, facilitate prototyping, training, and performance monitoring, leading to cost savings and improved productivity. AR-based remote assistance became crucial during the COVID-19 pandemic, allowing experts to provide real-time guidance to on-site workers.

Training and Collaboration: Immersive training with XR can enhance retention and practical skills, while XR guidance provides step-by-step instructions, reducing errors and improving productivity. AR enables real-time, 3D collaboration on digital assets.

CHAPTER 5   KEY USE CASES FOR XR

Questions

- What are some examples of how VR can be used for simulations in a business context?

- What are the main findings of the survey conducted by AiShed and AIShed regarding XR training preferences among engineers?

- Are there any limitations or challenges when implementing AR for collaboration?

CHAPTER 6

Deployment

Technology is best when it brings people together.

—Matt Mullenweg

The chapter discusses the deployment and implications of XR (Extended Reality) technologies in the Industrial Metaverse. It highlights the benefits, challenges, and best practices associated with XR deployment. The document covers health and safety concerns, such as symptoms like nausea and headaches, and the potential for sleep disorders and epilepsy attacks. It also emphasizes the importance of a systematic approach to deployment, including requirement validation, testing, and configuration control.

The chapter outlines the operational benefits of XR, such as enhanced performance through immersive experiences, virtual training modules, and AR-assisted maintenance tools. It also discusses the importance of time synchronization between multiple Metaverses for accurate data simulation and the need for robust security and privacy measures.

Additionally, the chapter explores the concept of the Metaverse, its potential applications in various fields, and the importance of interoperability, interactivity, synchronization, digital continuity, immersion, inclusion, and integration. It emphasizes the need for a model-based approach to enhance simulations and the importance of establishing a single source of truth for data.

Critical is the role of intrapreneurship in driving innovation within companies and the importance of creating a community of XR champions to facilitate successful organizational change initiatives. It concludes by

CHAPTER 6 DEPLOYMENT

discussing the potential return on investment (ROI) of XR technologies and the need for continuous monitoring and updating of XR systems to ensure optimal performance.

Deploying in the Metaverse

The deployment of XR in the Industrial Metaverse requires a multipronged approach. In general, the deployment process is an essential part of the software development process. The objective of such a process is taking a software from development to production. The activities to be performed in the deployment process are mainly ensuring that the validated requirements are verified. In fact, this process shall involve testing, configuration control, and trial. Based on the above, when an XR deployment is required for realizing an Industrial Metaverse, companies should invest in creating immersive XR experiences, such as virtual training modules, to improve workforce efficiency. They should implement AR-assisted maintenance tools to reduce equipment downtime and enhance safety.

The current era is indeed filled with exciting possibilities as companies strive to become pioneers in implementing an Industrial Metaverse. By adopting these strategies, companies can leverage the benefits of XR to enhance their operational performance, leading to increased productivity and profitability. This virtual and structured universe, developed in-house, offers immense potential for marketing and industrial production.

So, why may Industrial Metaverse be realized in your company? The Metaverse can help companies create new products, services, and experiences that are more engaging and intuitive than ever before. Additionally, companies can quickly and cost-effectively develop and deploy new solutions and services by leveraging blockchain technology. In fact, Metaverse's decentralized network allows businesses to access data and resources from anywhere in the world, eliminating the need for expensive infrastructure and personnel. For example, XR, as a part of Metaverse, can be used to create a digital twin of physical assets, enabling remote monitoring and predictive maintenance activities.

In the next chapter, some examples from industry will be treated in order to provide readers with a sort of inspiration that can help businesses, reduce costs, and increase efficiency.

Examples from Industry

To better understand why Industrial Metaverse can be introduced in your company in order to improve the activities' performance or increased productivity and profitability, we can see, for example, how the usage of XR technology enhances the overall customer experience in a highly competitive market, differentiating a business from its competitors.
XR technology can revolutionize the industry by providing immersive experiences for customers, employees, and investors even before they are put on board. By creating 360-degree virtual tours of the factories, they can offer a better understanding of what the company is manufacturing or how the company is working to realize a certain product, allowing customers to learn more about a product or service in a more engaging way. For example, XR technology can be used to create virtual showrooms, allowing customers to explore products in a realistic 3D environment. This can be used to provide customers with a more interactive and engaging shopping experience. On the other hand, companies can use XR technology to optimize their own needs or to make the best decision, for example, airlines, cabs, and train companies can use XR to showcase different destinations and offer travelers a preview of their upcoming trip.

Companies can also use XR technology to create virtual training simulations, allowing employees to learn new skills in a safe and immersive environment, sharing verbal information by virtual conferences and meetings. This can cause confusion because this type of interaction is not away from the current status of most companies, but it adds to acoustic experience an immersive experience which is made by a virtual environment where employees can move in a more engaging and efficient way.

CHAPTER 6   DEPLOYMENT

Metaverse provides a secure and efficient platform for businesses to store and transfer data, as well as to create and manage digital assets. This eliminates the need for costly third-party intermediaries, such as banks and other financial institutions, to facilitate transactions. Additionally, Metaverse's smart contracts allow businesses to automate certain processes, such as payments and contracts, which can reduce the need for manual labor and associated costs.

It all sounds so simple, but unfortunately, to realize the deployment of XR, or more generally to introduce the Metaverse in our company, we need to follow some rules.

We saw how the System Engineering approach can help to systematically identify objectives for XR systems and define requirements, constraints, and interfaces (see interviews in Chapter 7). System Engineering can help to deliver XR systems that meet the needs and expectations of users and customers, which is the objective of a company. System Engineers can collect all of the information necessary for configuration control. By breaking down the system into components and analyzing their interactions, their approach can ensure that the XR system functions meet the desired performance metrics.

Moreover, the approach can help to manage and optimize the design process, coordinate with stakeholders, and assess the system's life cycle costs and risks. Embracing this cutting-edge solution has the potential to revolutionize the industry and create a truly unique experience for both businesses and customers alike.

What the Metaverse Is and Is Not About

The concept of the Metaverse has been closely associated with various technologies such as digital twin, cyber physical infrastructure, computer vision, and more. However, it is important to note that the Metaverse is not about replacing real life or gamifying industrial processes, but rather it is a platform that allows for the representation of reality in a way that is both accurate and engaging. It is often difficult to determine the accuracy of a representation of reality, as it is often subjective and based on individual perspectives. Therefore, as we mentioned in previous chapters, who transposes the real word in a virtual environment should associate to virtual representation a certain degree of truth. The degree of truth can be affected by the context in which the representation is made. For example, if a representation of reality is based on inaccurate or incomplete information, then the degree of truth may be lower than if the representation was based on accurate and complete information. By an established degree of truth, the Metaverse can be used to

- Create a digital representation of the physical world, allowing for a more immersive experience

- Interact with each other and with the environment in a way that is more realistic than what is possible in the real world

- Explore and experience new things, and where people can create their own stories and experiences

- Connect people in ways that are not possible in the physical world

Therefore, the Metaverse can be seen as a new avenue that offers opportunities for enhanced data distribution, digital assets, and digital continuity. It goes beyond mere entertainment or virtual experiences and has potential applications in various fields. By leveraging infrastructure and technological advancements, the metaverse enables users to immerse themselves in a digital realm where they can interact with others, explore virtual environments, conduct business transactions, and create unique experiences.

CHAPTER 6　DEPLOYMENT

The Metaverse is indeed being touted as a potential evolution of the Internet, offering a more integrated and immersive way for businesses to connect with their customers. It has the potential to streamline interactions, reducing bureaucratic hurdles and creating seamless experiences. This transition reflects a shift from traditional document-based systems engineering toward a model-based approach, where virtual environments can be built and customized to meet specific needs. By embracing the metaverse, businesses can potentially enhance their online presence and create innovative ways to engage with their target audience.

The concept of an industrial metaverse, though still evolving, is generally expected to incorporate several key elements. These include interoperability, interactivity, synchronization, digital continuity, immersion, inclusion, and integration. Let's explore each of these in more detail:

1. Interoperability: The industrial metaverse will likely emphasize the ability for various systems and technologies to seamlessly communicate and integrate with one another. This will enable efficient collaboration and data exchange across different platforms.

2. Interactivity: A crucial aspect of the industrial metaverse will be its focus on enabling real-time interaction between users and virtual environments or objects. This can involve immersive experiences through augmented reality (AR) or virtual reality (VR), allowing users to engage with digital content in a more tangible way.

3. Synchronization: In order to create a cohesive metaverse experience, synchronization plays a vital role. It ensures that different devices or platforms are able to maintain consistent data updates in real time across the entire ecosystem.

4. Digital continuity: The concept of digital continuity involves maintaining a seamless flow of information throughout various stages or processes within the industrial metaverse environment. This helps ensure that data remains accurate and up-to-date throughout its life cycle.

5. Immersion: Encouraging deep engagement within the virtual environment is another important aspect of the industrial metaverse experience. By providing realistic graphics, haptic feedback systems, spatial audio cues, or other sensory inputs, users can feel fully immersed in their interactions.

6. Inclusion: The goal of inclusivity in the industrial metaverse is to provide equal access for all individuals regardless of their abilities or backgrounds. This means developing user interfaces that are intuitive for everyone and designing experiences that accommodate diverse needs.

7. Integration: Integration refers to how well different components come together within the overall framework of the industrial metaverse ecosystem. This includes integrating hardware devices (such as sensors or wearables) with software applications seamlessly so they can work together effectively. As the concept of the industrial metaverse continues to evolve, these elements will shape its development and determine how users engage with this immersive digital environment.

CHAPTER 6   DEPLOYMENT

The industrial metaverse is a concept that goes beyond entertainment and digital facsimiles of ourselves. While it may involve elements such as NFTs, XR gadgetry, and even aspects of cryptocurrency as part of the evolving Web 3.0 ecosystem, its focus is not solely on these aspects. Unlike a virtual world in the style of "Ready Player One" or an escape from reality, the industrial metaverse aims to create immersive digital environments that serve practical purposes in various industries. It seeks to facilitate collaboration, innovation, and problem-solving by leveraging technology to enhance productivity and efficiency. Rather than being centered around virtual tourism or entertainment-driven experiences, the industrial metaverse strives to transform industries such as manufacturing, healthcare, education, architecture/design, retail, and many more. It aims to provide tangible benefits by enabling remote work setups, advanced simulations, and training programs for professionals across different fields. In summary, while there may be overlaps with certain technologies or concepts related to the entertainment-focused metaverse narrative seen in popular culture today (such as NFTs or XR gadgets), the industrial metaverse focuses on practical applications within real-world industries rather than escapism or virtual tourism.

How to Harness the Potential of MBSE

Extended Reality (XR) has the potential to revolutionize systems engineering approaches by enabling truly model-centric systems. By utilizing a single system model as the "source of truth" for all embedded system analyzing tools, XR can facilitate a seamless flow of information between the system model and virtual reality environments. This allows systems engineers to conduct interactive immersive simulations of scenarios described in the system model and analyze system performance. On the other hand, incorporating XR into MBSE (Model-Based Systems Engineering) is its ability to enhance requirements analysis. By involving customers early in the product life cycle through XR environments, requirements can be analyzed more efficiently. Customers can step into an XR environment and visualize how the system design would look like, as well as interact with it. This interactive experience provides valuable insights that can inform modifications to product or service specifications and design. With XR, systems engineers have an unprecedented opportunity to engage customers in a more tangible way during the development process. By immersing customers in virtual environments where they can interact with simulated designs, engineers can gather feedback and make informed decisions based on user experiences. In summary, integrating Extended Reality into MBSE approaches holds immense potential for enhancing systems engineering processes. The ability to leverage immersive simulations and customer interactions within XR environments enables more efficient requirements analysis, leading to improved product/service specifications and design iterations. Embracing this technology will undoubtedly contribute toward creating truly model-centric systems engineering approaches for future projects.

Create a Community of XR Champions Within Your Organization

The Change Champion Network is an invaluable group of employees dedicated to facilitating successful organizational change initiatives. By acting as a liaison between peers and leadership, they help foster buy-in for and minimize resistance to change (Figure 6-1). When it comes to introducing XR (Extended Reality) technology in your organization, it's crucial to create a comprehensive evangelization strategy that encompasses both top-down and bottom-up approaches.

CHAPTER 6 DEPLOYMENT

Figure 6-1. *The role of a change champion*

This ensures that everyone is engaged and excited about the potential benefits of XR. To kickstart this process, consider gathering all individuals who are already interested in XR technology or those who can be encouraged to develop an interest by organizing engaging events. XR, with its strong connection to gaming, presents an ideal opportunity for organizations looking to embrace technological advancements. By highlighting the immersive and interactive nature of XR technology through informative sessions, demonstrations, workshops, or even gaming competitions within your organization, you can generate curiosity and build enthusiasm among employees at all levels. Additionally, involving influential leaders who are supportive of XR can add credibility and encourage broader adoption throughout the organization. When employees witness their managers showcasing enthusiasm for this innovative technology, it helps reinforce the notion that embracing XR is not just a passing trend but a strategic investment in the future. Remember that communication plays a vital role throughout this process. Regularly sharing success stories from early adopters or showcasing how other companies have benefited from integrating XR can inspire others within your organization as well. Overall, by leveraging the Change Champion Network alongside creative evangelization efforts focused on both top-down support and bottom-up engagement through various activities related to gaming-based experiences offered by XR technology, you'll be well positioned for success in implementing this exciting new initiative within your organization.

Encourage Intrapreneurship

Intrapreneurship is vital for driving innovation within a company. Encouraging intrapreneurship involves creating an environment that fosters creativity, risk-taking, and experimentation (Figure 6-2). Companies can provide resources such as time, funding, and training for employees to work on projects that are outside of their usual job duties. Leaders can also set aside specific time for brainstorming and idea generation. Recognizing and rewarding employees who take initiative and contribute new ideas will ultimately lead to a more innovative and successful organization.

CHAPTER 6   DEPLOYMENT

Figure 6-2. *Illustration of technology adoption business impact*

A powerful approach to driving change is through collective action. This involves rallying individuals and groups that share a common goal and empowering them to work together toward a shared vision. It requires effective communication, collaboration, and a willingness to act on the feedback and insights received. By harnessing the efforts of many, collective action can create a groundswell of momentum that is difficult to ignore, thereby increasing the likelihood of successful change. This approach has been used successfully in a variety of settings, including social justice movements, environmental activism, and corporate transformation efforts.

In particular, when it comes to emerging technologies like XR, embracing intrapreneurship becomes even more important. The Industrial Metaverse, a digital universe where physical and virtual worlds converge, offers immense potential for industries across the board – from manufacturing to healthcare.

By fostering an environment that encourages intrapreneurship in XR technology, companies can tap into the transformative power of immersive experiences. This allows employees to think outside the box and explore new possibilities for their respective industries.

By creating a culture that values intrapreneurial thinking in relation to XR technology adoption, businesses can unlock new opportunities for growth and innovation. Encouraging employees to step forward with their ideas outside of traditional boundaries will foster collaboration across departments and drive progress toward achieving organizational goals.

In conclusion, embracing intrapreneurship when it comes to emerging technologies like XR is essential for companies aiming to stay competitive in today's rapidly changing world. By empowering employees and creative individuals beyond ROI calculations and siloed thinking, organizations can harness the true potential of XR technology while driving meaningful change within their industry.

How to Realize the Full Potential of XR

The deployment process of XR in the complex Industrial Metaverse is a multi-faceted endeavor that requires careful planning and execution. In general, it is important to ensure that the XR solution is properly configured and integrated into the existing environment; this is the main duty of organizations since XR shall be deployed successfully and be ready to use in the complex Industrial Metaverse.

This process involves a variety of steps, from the initial assessment of the environment to the implementation of the best XR solution for the specific use case, including the hardware and software requirements, the user experience, and the security and safety considerations. Once the best solution is identified, the next step is to develop a plan for the deployment, integrating the XR solution into the existing environment. The final step is to test and validate the XR solution to ensure it meets the requirements and is ready for deployment.

CHAPTER 6   DEPLOYMENT

In order to fully realize the potential of XR, there are certain limitations that need to be addressed. One crucial aspect is the control of Metaverse solutions. By exerting control over the development process, we can ensure that these solutions are adaptable to more challenging scenarios, particularly in terms of security and privacy.

To effectively harness the power of the Metaverse, it is essential to establish a single source of truth for data. This means having a centralized repository or data lake that serves as the authoritative reference for all information within the metaverse environment. This allows for seamless integration and synchronization across different platforms and applications. In fact, the main challenge of engineers is to realize the linkage between two or more Metaverses. This is an exciting one, as it allows for a much more immersive and expansive virtual world experience. In this type of virtual world, users can explore and interact with a variety of different Metaverses, each with its own unique environment and features. These Metaverses can be connected to the main Metaverse, allowing users to move between them and explore the different areas. This type of virtual world also allows for a much more realistic simulation, as the different Metaverses can be linked together and interact with each other.

Furthermore, adopting a model-based approach can enhance simulations within the metaverse. By creating digital models that accurately represent physical objects or systems, we can simulate various scenarios and evaluate their outcomes before implementing them in real life (see interview of Dr Laughlin in Chapter 7). This not only saves time but also reduces costs and minimizes risks associated with physical prototyping.

Another critical aspect is ensuring digital continuity within the metaverse environment. This involves maintaining consistency across different stages of development and ensuring compatibility between different software tools and technologies used within the ecosystem. By establishing robust standards and protocols, we can ensure smooth transitions between different components of the Metaverse.

However, realizing this full vision comes with challenges related to security and privacy. As more sensitive data is generated within the metaverse environment, it becomes crucial to implement robust security measures to protect against unauthorized access or data breaches. Additionally, privacy concerns must be addressed by implementing transparent policies regarding data collection, usage, and sharing within the metaverse ecosystem.

The problem of **time synchronization** between two or more Metaverses is considered when the organization's purpose is aimed to simulate a lot of data in a timely manner. We suppose a user could explore virtual maintenance activities in one Metaverse, and then move to a virtual logistic in another Metaverse, and the two environments could interact with each other, for example, in the accounting office. A user can explore different environments and interact with them in a much more realistic way. There is a need for time synchronization between them. However, how can multiple Metaverses be linked together? This is necessary to ensure that events in one Metaverse are accurately reflected in the other Metaverses (i.e., maintenance outcomes reflect purchasing components up to shipping and tracking). Without proper time synchronization, users may experience lag or other issues when interacting with other users or objects in the Metaverse. Potential solutions to ensure that all users in the Metaverse are experiencing the same time are the usage of

- A distributed **time synchronization protocol (blockchain-based time synchronization protocol)**. This protocol would allow each Metaverse to maintain its own time, while also synchronizing with the other Metaverses. A signal is required to synchronize all Metaverse clocks. This would involve each Metaverse sending out its own time signal, which would be received by the other Metaverses. This would require the Metaverses to have a reliable network connection, but would be more secure than using a centralized time server.

- A centralized **time synchronization server**. This server would be responsible for maintaining the time in all Metaverses, in order to accurately reflect all events in each Metaverse. This server would be responsible for keeping track of the time on each Metaverse to communicate with the time server, which could be done through a secure protocol such as HTTPS.

Note XR technologies can also be used for time synchronization between multiple Metaverses. XR technologies such as augmented reality and virtual reality can be used to create a shared virtual environment, where all Metaverses are running in the same virtual space. This would allow the Metaverses to synchronize their clocks more easily, as they would all be running in the same virtual environment.

Another potential solution may have a cultural aspect related to being aware of the situation. By being aware of the environment, you can better plan and coordinate activities to ensure that everyone is on the same page and that tasks are completed in a timely manner. This is especially important in a business setting, where time is of the essence and deadlines must be met. It is the ability to be aware of what is happening around you and to be able to anticipate potential problems before they occur. With situational awareness, you can identify potential problems before they arise and take steps to address them.

In conclusion, there are several potential solutions to the problem of time synchronization between multiple Metaverses. These solutions range from using a centralized time server to using a distributed time synchronization protocol, to using XR technologies to create a shared virtual environment. Each of these solutions has its own advantages and disadvantages, and developers should consider which solution is best suited to their needs.

Return On Investment (ROI)

Despite being a relatively new field, XR offers numerous benefits such as improved training, enhanced customer experiences, and increased efficiency, which can have a positive impact on the bottom line. The ROI for XR can be measured in terms of cost savings, productivity gains, increased sales, and improved customer satisfaction. As more businesses adopt XR technologies, the ROI is expected to continue to grow.

Productivity improvements are the process time savings that users and people supporting or interacting with them experience from the use of the XR solution. This translates to more efficient and effective interactions, optimizing workflows, and improving overall productivity levels of businesses and organizations.

When calculating the percentage of time saved, you must be sure to calculate the total amount of time spent on the legacy process that is being replaced, and subtract that from the amount of time required to complete the process using the XR solution. Make sure to consider hidden costs like initial and on-going training for use of the solution, as well as training required for new hires who will utilize the solution. Also consider costs related to initial, on-going and new hire training in secondary business processes that are integrated with the XR solution (e.g., order processing, sales and marketing, etc.).

Whether or not your activity is ready to jump in the Metaverse, there are some key questions that need to be answered to create a meaningful business case for an XR application.

- How many employees will use the new solution?
- What is the amount of time (minutes) that each user will save per day from using the AR solution?
- How many managers will realize productivity improvements (process time savings) from the use of the new solution?

- What is the amount of time (minutes) that each manager will save per day from using the AR solution?
- How many other employees will realize productivity improvements (process time savings) from the use of the new solution?
- What are the annual maintenance costs on hardware that will be eliminated by using the AR solution?

The complete set of questions is covered in the AREA Online ROI calculator that we recommend using when preparing a pitch. Nevertheless, it is recommended to tailor the questions for the specific operational environment within your organization.

A Proposed Deployment Process

In the Industrial Metaverse, an XR deployment process shall be introduced. In general, such a process can involve several steps, beginning with assessing a company's needs and determining the appropriate technology to use. A team of experts is required for creating content to justify the XR applications. Once the adoption of XR (Extended Reality) technology is approved, the next steps involve the integration of both hardware and software components. This is followed by systematic testing to ensure that the functional requirements are intended. After successful testing, the XR system is deployed across relevant departments, and comprehensive training is provided to employees to familiarize them with its use. Continuous monitoring and periodic updates are crucial for sustaining the performance of XR (Extended Reality) systems. A proactive strategy is essential to guarantee that such technology adapts to the changing requirements optimizing its overall benefits.

Deploying XR use cases in an organization requires a systematic approach that can vary depending on the organization's size and specific needs. While not all steps may be necessary for every organization, it is essential to have an awareness of the process and adapt it to ensure efficient deployment.

CHAPTER 6 DEPLOYMENT

The deployment process of XR in the Industrial Metaverse is a complex process that requires careful planning and execution. Following a systematic approach which is familiar to systems engineers, the leader of such an activity shall tailor the following milestones:

- To define the scope of the project and the objectives to achieve
- To define stakeholders' requirements, including business needs
- To define constraints and domain of the project
- To collect the technologies and resources available, even introducing all of the technical specifications and drawings
- To analyze the ROI, both for full deployment and POC (Proof of Concept)
- To develop a plan for implementing the XR solution, including the selection of the appropriate XR platform
- To develop the integration of the XR platform into the existing environment
- To test and verify the XR solution
- To validate the entire system that represents the Industrial Metaverse

A special mention shall be made for the plan development. The team leader (or the team of champions) must introduce

- The installation of the XR platform. This includes the drawing and technical specification before collecting.
- The configuration of the XR platform. This includes the XR integration sketch with the existing systems.

- The compliance with industry standards and regulations of the XR solutions. This also includes the healthcare and safety protocols that would be set up ad hoc.

- The reliability and maintainability process for the XR solutions. This includes the development of a maintenance schedule, the implementation of system updates, and the monitoring of the system.

- The training plan, including a syllabus. This exposes and teaches the users how to use the technology and the application.

Note XR Benchmarks are recommended to capture the various technologies and to assess their relevance in the industrial context. We suggest collecting use cases to understand business needs, ideas, and pain points for understanding the potential impact of the XR solutions on the organization. Potential users of collected use cases shall be defined in order to assess their preferences and needs. This will help to gain insights on how to best design and develop the XR solutions to meet the users' needs and expectations. This will enable them to minimize costs and times and to ensure that XR solution is developed in the most effective and efficient manner.

Once the system is operational, the team of champions must ensure that the system is secure, that is the implementation of security protocols, the monitoring of the system, and the implementation of updates and patches are verified. On the other hand, the team must develop a plan to ensure the system is always up-to-date with the latest technologies and trends.

CHAPTER 6   DEPLOYMENT

Finally, the team must develop a plan to ensure the system is optimized for the Industrial Metaverse. This includes the development of a strategy to ensure the system is optimized for the use cases and applications that are being deployed in the Industrial Metaverse. Additionally, the team must develop a plan to ensure the system is integrated with other systems and platforms in the Industrial Metaverse. This includes the integration of the XR platform with other applications, with other systems, and finally with the Industrial Metaverse.

Before definitely deploying XR project in the industrial environment, the team shall present to their organization

1. The scope of the project
2. Demos and exchanges in situ, including prototype and development of a stand-alone solution (a vision of the solution which will be developed is recommended)
3. The integration of the solution as a standard equipment process in the business
4. Business impacts. The assessment of the cost and time required to implement the XR solutions and the potential return on investment (ROI)
5. Benefits over time, once the solution is in place
6. Established KPIs enabling to validate the hypothesis in a representative environment
7. The development of a more advanced version of the prototype (MVP)
8. The delivery of an MVP, which can be used in stand-alone mode

CHAPTER 6   DEPLOYMENT

9. The certification plan of the XR solution according to company standards
10. Contacts all other potential customers to propose the solution

Do Not Reinvent the Virtual Wheel!

When introducing untested hardware into your organization, it is indeed crucial to exercise caution when it comes to using untested software. To ensure a smooth integration of novel and highly refined devices, one key recommendation is to explore off-the-shelf applications or applications that are already functioning well within your organization. By leveraging existing software, you can focus on adapting and optimizing them for this new hardware platform. Using off-the-shelf applications that are already proven and reliable can offer several advantages. Firstly, these applications have typically undergone rigorous testing processes or have been used by other organizations successfully. This minimizes the risk of encountering major compatibility issues or technical glitches when integrating them with new hardware. Additionally, opting for existing software allows you to tap into a pool of knowledge and resources available through user communities or support channels. This can be valuable in troubleshooting any challenges that may arise during the adaptation process. Evaluate whether it offers features relevant to your organization's unique needs and workflows as well. Once you've identified an appropriate application, work on customizing (tailoring) it for optimal performance with the new hardware platform. Collaborate closely with your IT team or developers who have experience in adapting software for specific devices. Conduct thorough testing to ensure seamless integration and identify any areas where further adjustments may be needed. By following these recommendations, you can minimize risks associated with using untested software while ensuring a smoother integration process for new hardware within your organization.

Cybersickness

As organizations increasingly integrate XR technology into their operations, it is becoming more common for individuals to be exposed to immersive experiences. However, as with any mass adoption, some people may be more susceptible than others to the potential side effects of XR technology. One such side effect that companies are actively addressing is cybersickness or VR sickness. Cybersickness refers to the discomfort and nausea that some individuals experience when using virtual reality (VR) or augmented reality (AR) devices. This condition arises due to a discrepancy between the visual information received through these devices and the body's sensory cues, leading to feelings of dizziness and disorientation. Recognizing this issue, organizations have taken significant measures to address cybersickness in order to ensure a comfortable user experience for all individuals. By investing in research and development efforts, companies are continuously refining XR technologies and implementing strategies that mitigate these unpleasant symptoms. Technological advancements such as improved tracking systems, reduced latency, higher display resolutions, and enhanced motion controllers have significantly contributed toward minimizing cybersickness (Figure 6-3).

CHAPTER 6   DEPLOYMENT

Figure 6-3. *Abstract illustration of cybersickness (Source: DALL-E 2)*

Additionally, software developers are incorporating techniques like adaptive rendering algorithms and smoother locomotion mechanics into applications designed for XR experiences. Furthermore, comprehensive user education programs have been implemented by companies intending to deploy XR technology at scale. These programs aim to familiarize users with best practices for avoiding cybersickness by gradually acclimating them to virtual environments through controlled exposure sessions. By prioritizing user comfort and consistently refining their products based on user feedback and scientific insights from ongoing studies on cybersickness mitigation techniques, organizations demonstrate their commitment toward ensuring a positive immersive experience for everyone involved. In conclusion, while some individuals may be more susceptible than others to cybersickness when using XR technology at

CHAPTER 6   DEPLOYMENT

scale in large or small organizations alike; companies are proactively working toward minimizing this side effect through technological advancements and comprehensive user education programs.

> The use of XR technologies may lead to the onset of the following symptoms: pallor, feeling unwell, disorientation, headaches, sweating, fatigue, nausea, vomiting, tachycardia, hypersalivation, balance disorder.

Recommendations for Safe XR Demonstration Sessions

There is a lack of research on the effects of XR technology among a widespread population from young to adults. In fact, while XR technology has been shown to have prosocial benefits for younger users, it is unclear if these benefits extend to older adults. As such, further research is needed to understand the potential benefits and drawbacks of XR technology for older adults.

We saw that XR is an innovative solution that can improve industrial activities while considering the well-being of employees and other users. However, it is important to consider the impact on cognitive and emotional well-being, as well as other psychosocial factors, in addition to focusing solely on employees. Factors such as improved communication, training, and collaboration can positively influence operational efficiency and effectiveness. XR technology can also enhance safety measures and reduce the risk of accidents.

In order to ensure safe use of XR in the workplace, clear rules should be developed in consultation with the occupational health department since it is important to prioritize the health and well-being of employees.

CHAPTER 6   DEPLOYMENT

It is recommended to develop clear and agreed rules for safe use of XR in the workplace. These rules should take into account expected exposure time and conditions.

In summary, the following are some recommendations:

- Stop the XR activity when symptoms of cybersickness occur.
- Start with a short immersion with gradual increases.
- Avoid driving and use of hazardous machinery for one hour following the immersion.
- Do not use XR in the two hours preceding bedtime.
- Prefer seated experiences whenever possible.
- Make sure there is a suitable place to take a break after the exposure.

Some companies recommend the compilation of a Virtual Reality Sickness Questionnaire to self-assess the staff after the experiences, especially for first-time and occasional users.

Preparation of the XR Experience

Research shows that product demos are a crucial and transformative part of the purchase journey for VR devices. Using XR requires expertise and a thorough approach for industrial purposes to ensure safety.

The core of the XR design-and-development process mainly consists of story-boarding, writing scenarios, and testing. The final product should meet the desired outcomes, be engaging and entertaining, and easy to use.

Before experiencing XR, it is important to ensure that the device is charged and functional and to identify the learning objective, tone, and target audience to curate appropriate content. The user should also familiarize themselves with the XR application or software that will be used and ensure that it is properly installed on the device. It is recommended to be in a well-lit and spacious room with minimal obstacles to avoid accidents during the XR experience since it is important to have a clear understanding of any safety warnings or precautions related to the specific XR experience. Finally, the user should make sure they are in a comfortable position to avoid discomfort or injury during the experience (Figure 6-4).

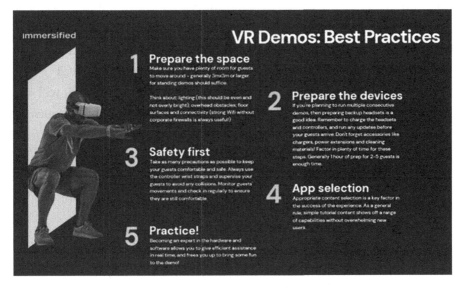

Figure 6-4. Best practices to organize a VR Demo (Source: Immersified)

Cybersecurity

XR is a part of the Internet of Things (IoT) and requires strong cybersecurity measures while immersive and realistic interaction features of XR also raise concerns (Figure 6-5).

CHAPTER 6   DEPLOYMENT

Nowadays, the benefits of robust company-wide cybersecurity systems are quite obvious and a number of measures are already embedded in the ICT infrastructure of many companies. Nevertheless, the introduction of XR technology brings new challenges. For instance, it is quite common to first introduce XR in a company from the R&T or R&D door. New digital technology is often incompatible with the cybersecurity measures already in place in the organization. If you are a bit tech-minded, you probably heard people involved in testing immersive technology talk about how they need to set up bespoke, non-standard systems to perform their testing. The use of Internet connections that are separate from the company network is not uncommon, as well as the use occasionally of consumer solutions to bypass the restriction of the company ICT infrastructure. That is a huge paradox; to test new technology that will benefit the company, people need to actively bypass the security measures in place to introduce that same technology, till full compatibility can be achieved.

CHAPTER 6 DEPLOYMENT

Figure 6-5. *Abstract illustration of a Troy horse (Source: DALL-E 2)*

Sometimes compatibility is not achieved as some devices require a constant bidirectional data stream with the manufacturer to work correctly. Often this feature could be omitted from the specifications of the device. As XR devices are by nature capable of acquiring multisensorial information from the environment, they could be seen as the perfect Troy horse from a cybersecurity perspective. The ability to collect such detailed data raises questions about privacy and security. To mitigate these risks, it becomes crucial to thoroughly research and consider factors such as device origin, manufacturer reputation, encryption protocols used for data transmission, and any known vulnerabilities. The initial stage for effectively contrasting XR security and privacy risks is to comprehensively determine the method of implementation of the new XR tools. This could involve exploring multiple avenues, each presenting distinct challenges. For instance, if one plans on using Augmented and Mixed Reality technologies to develop their XR solutions, it is important to understand the potential risks associated with data sharing and collection.

The XR world presents different security and privacy concerns for various companies depending on their industry and compliance regulations. For example, the financial or healthcare industries may require specific guidelines to be met. It is essential for companies to understand their compliance obligations and implement appropriate measures to protect user data. A holistic approach to security and privacy will be essential in gaining user trust and ensuring the long-term success of XR technology.

Additionally, staying informed about software updates and patches can help address potential security issues. It is essential to stay vigilant when choosing XR devices or any other technology that connects to networks or collects personal information. Taking proactive steps toward securing these devices can help ensure a safer user experience in this evolving technological landscape.

As companies delve into security and privacy measures, safeguarding the massive amounts of data collected through these tools is the top priority. The XR environments are rich with sensitive data important for businesses, ranging from user environment scans to eye-tracking data. Protecting this data involves implementing secure network infrastructure, advanced encryption, access controls, and policies that regulate data sharing. Furthermore, companies should invest in educating their staff on security threats and the importance of ethics in data handling, as they aim to ensure the privacy and protection of personal information.

A detailed assessment shall be carried out with regard to the attack channel, potential cyber threats, and prospective attacks that could impact XR applications. This analysis involves a review of a wide range of sources, including research publications, conferences, online content, and expert viewpoints. Through this process, potential security vulnerabilities can be identified, which will need to be addressed in order to enhance the safety and security of XR applications.

Dealing with Unions and Work Councils

We conclude the chapter with an often neglected aspect of the introduction of XR technology in the workplace. The introduction of XR in the workplace brings various considerations, including an often neglected aspect related to personal data management. In addition to the healthcare concerns mentioned earlier, unions and work councils are increasingly concerned with the handling of personal data in industry, both from software and hardware perspectives. Ensuring that personal data is managed properly is crucial in maintaining privacy and complying with regulations. Organizations implementing XR technology must address these concerns by implementing robust data protection measures. This involves establishing clear policies and procedures for collecting, storing, processing, and disposing of personal data. From a software standpoint, organizations should prioritize secure coding practices to minimize vulnerabilities that could lead to unauthorized access or data breaches. Regular security audits and updates should be conducted to stay ahead of potential threats. On the hardware side, organizations need to ensure that XR devices are designed with privacy in mind. This may include features such as user authentication controls, encryption capabilities, and secure storage of sensitive information. Collaboration between employers, employees' representatives such as unions or work councils, and relevant regulatory bodies is crucial to address these concerns effectively. By working together proactively, organizations can strike a balance between reaping the benefits of XR technology and safeguarding individuals' rights to privacy.

For instance, in Germany, it is important for companies to obtain approval from the work council before implementing any software or hardware used by employees. Failure to do so could result in severe sanctions. When it comes to VR training, mandating training exclusively in virtual reality may be seen as discriminatory toward employees with low tolerance for this technology. Therefore, an alternative "traditional" solution must always be provided alongside VR training. This requirement currently limits the application of VR training in an industrial metaverse unless it can be accessed through different means. It is crucial for companies operating in Germany to navigate these regulations and ensure that all necessary approvals are obtained while providing inclusive solutions that accommodate the varying needs and preferences of their employees.

In summary, managing personal data is an important consideration when introducing XR technology in the workplace. Addressing these concerns through robust policies and procedures for software and hardware management will help maintain privacy standards while embracing the potential benefits of XR technology.

Takeouts

- **Deployment Process**: The deployment of XR in the Industrial Metaverse involves multiple steps such as requirement validation, testing, configuration control, and trial, aiming to improve workforce efficiency and safety.

- **Operational Benefits**: XR can enhance operational performance by creating immersive experiences, virtual training modules, and AR-assisted maintenance tools, leading to increased productivity and profitability.

- **Health and Safety Concerns**: XR technologies can cause symptoms like nausea, headaches, and disorientation, and may also lead to sleep disorders and epilepsy attacks in susceptible individuals.

- **System Engineering Approach**: Using a System Engineering approach helps in systematically identifying objectives, defining requirements, and optimizing the design process for effective XR deployment.

- **Intrapreneurship and Innovation**: Encouraging intrapreneurship can drive innovation within companies, allowing employees to explore new possibilities with XR technology and contribute to organizational growth.

Questions

- How can immersive XR experiences like virtual training modules benefit the workforce efficiency?
- What are some examples of industries that can benefit from utilizing the Metaverse?
- How can organizations create a community of XR champions to implement XR effectively?

CHAPTER 6 DEPLOYMENT

- How can companies ensure the proper configuration and integration of XR deployment processes?

- What are some key milestones to consider when defining project scope for XR deployment?

- What are some psychosocial factors that should be considered when implementing XR technology at work?

CHAPTER 7

Learning from Experience

This chapter explores the convergence of Extended Reality (XR) technologies and Model-Based Systems Engineering (MBSE) in industrial applications, featuring insights from experts in the field. Several interviews are summarized here.

Dr. Brian Laughlin from Boeing discusses the integration of XR technologies with MBSE, highlighting the concept of the "Industrial Metaverse" and its potential to revolutionize industrial processes. Dr. Laughlin's early exposure to augmented reality at Boeing in the 1990s set the foundation for his career in leveraging advanced visualization techniques to enhance industrial processes. Dr. Laughlin emphasizes the benefits of VR for safe training and AR for reducing cognitive load by overlaying critical information onto the real world, enhancing production environments.

Additional interviews with experts Ryan Wheeler, Paul Davies, Francis Vu, Cedric Chane Ching, and Alexandre Mao provide diverse perspectives on XR's role in industrial applications, challenges, and future potential.

CHAPTER 7   LEARNING FROM EXPERIENCE

Interview with Brian Laughlin, PhD, TF (Boeing)

Figure 7-1. *Brian Laughlin*

The convergence of Extended Reality (XR) technologies and Model-Based Systems Engineering (MBSE) is paving the way for what experts are calling the "Industrial Metaverse." Dr. Brian Laughlin (Figure 7-1), a Technical Fellow at Boeing and a pioneer in the application of XR technologies in aerospace manufacturing is a leading expert in the deployment of XR solutions. With over 30 years of experience and a PhD in Human Factors Psychology, Dr. Laughlin offers unique insights into how XR and MBSE are revolutionizing industrial processes.

CHAPTER 7   LEARNING FROM EXPERIENCE

The Evolution of XR in Aerospace

Dr. Laughlin's journey with XR technologies began in the early 1990s when the term "augmented reality" was first coined at Boeing. "I was fortunate to be a very young process engineering analyst at the time, on the shop floor and in meetings with the guys who coined that term," he recalls. This early exposure set the stage for a career dedicated to leveraging advanced visualization techniques to enhance industrial processes.

One of Dr. Laughlin's guiding principles has been the concept of "information kidding" – bringing together critical information at the point of use and time of need for mechanics and inspectors. XR technologies have proven to be an ideal medium for this approach, allowing for advanced visualization of complex tasks and data.

Breakthrough Projects

Notable projects include Project Blackwell, which used Microsoft HoloLens for logistical asset visualization, and the Tanker Wiring Solution, which significantly improved aircraft wiring installation quality and efficiency.

Project Blackwell

Using Microsoft HoloLens, this project enabled the visualization and control of logistical assets in urban environments. Users could access detailed information about various assets through intuitive holographic interfaces.

Tanker Wiring Solution

Addressing challenges in aircraft wiring installation, Boeing implemented an AR solution that overlaid engineering drawings onto the physical aircraft. The results were remarkable:

- Over 90% improvement in quality
- Takt time reduced by approximately 50%

The Power of XR in Industrial Applications

Dr. Laughlin emphasizes the distinct advantages of both Virtual Reality (VR) and Augmented Reality (AR) in industrial settings:

Virtual Reality (VR):

- Enables safe training in potentially dangerous scenarios
- Allows for the development of muscle memory and procedural knowledge in a controlled environment

Augmented Reality (AR):

- More practical for production environments
- Reduces cognitive load by overlaying critical information directly onto the real world
- Eliminates the need for constant referencing between instructions and the physical workspace

The Industrial Metaverse: Bridging the Experience Gap

One of the most pressing challenges facing industries like aerospace is the growing experience gap caused by demographic shifts and retirement patterns. Dr. Laughlin sees the Industrial Metaverse as a powerful solution to this problem:

"XR can help to fill that gap. It's probably one of the most powerful uses of it. I can actually shore up, to some degree, the lack of experience and wisdom that comes with having years under your belt by doing a better job of explaining and visualizing complex tasks."

By embodying tribal knowledge in advanced visualizations, companies can rapidly accelerate the skill acquisition of novice workers, effectively "short-circuiting" the traditional learning curve.

Integrating XR with MBSE

The integration of XR technologies with Model-Based Systems Engineering (MBSE) represents a significant leap forward in industrial processes. Dr. Laughlin explains:

"We're applying MBSE through what we call visual work instructions. Instead of just telling you what to do, we can take the models we're using for reference and lay them out in an augmented reality fashion, digitally overlaying them onto the real world."

This integration offers several key benefits:

- Reduced cognitive load for workers
- Faster location and identification of specific components

CHAPTER 7   LEARNING FROM EXPERIENCE

- Seamless translation of digital models into physical space
- Improved accuracy and efficiency in complex assembly tasks

Challenges and Best Practices for XR Implementation

While the potential of XR in industrial settings is immense, Dr. Laughlin cautions against a technology-first approach. He offers several key insights for successful implementation:

- Focus on solving real problems, not just showcasing technology.
- Involve end users throughout the development process.
- Consider the entire ecosystem, including infrastructure, policies, and maintenance.
- Plan for scalability across different environments and user skill levels.
- Address potential barriers to adoption, such as device durability and user comfort.

Conclusion: The Future of the Industrial Metaverse

As XR technologies continue to evolve, Dr. Laughlin envisions a future where the interface between digital information and the physical world becomes even more seamless. He speculates about advancements in neural interfaces that could potentially bypass traditional display methods altogether.

CHAPTER 7   LEARNING FROM EXPERIENCE

The Industrial Metaverse, powered by the convergence of XR and MBSE, promises to revolutionize how we design, manufacture, and maintain complex systems. By providing contextually relevant information exactly when and where it's needed, these technologies have the potential to dramatically improve efficiency, quality, and worker performance across a wide range of industries.

As we move forward, the challenge will be to harness these powerful tools in ways that truly augment human capabilities, rather than simply adding to the information overload. With thoughtful implementation and a focus on real-world problem-solving, the Industrial Metaverse stands poised to usher in a new era of industrial innovation and productivity.

Interview with Ryan Wheeler (Collins Aerospace)

Figure 7-2. *Ryan Wheeler*

In the realm of XR technology and User Experience Design, Ryan has forged a path rooted in a solid educational foundation. Holding an associate degree in electronics, a bachelor's degree in electrical engineering technology, and master's degrees in business and systems engineering, he established himself during three years as a technician, a

CHAPTER 7   LEARNING FROM EXPERIENCE

formative experience that anchored his engineering perspective and paved the way for success in the visualization domain.

Transitioning from a technician to an engineer at Collins Aerospace (formerly Rockwell Collins), Ryan dedicated three years to manufacturing engineering, earning the prestigious Engineer of the Year distinction within an impressively short timeframe. In 2005, his trajectory led him to the advanced manufacturing organization, where he played a pivotal role in leading the invention of breakthrough technologies for competitive advantage. Through mergers and acquisitions, the organization evolved into RTX (formerly Raytheon Technologies), where Ryan currently contributes to the smart factory organization.

Ryan's proficiency in advanced visualization is deeply rooted in his innate talent for design. While excelling in this space, his aspiration is to pivot toward an autonomous systems role, capitalizing on skills not only in XR but, more crucially, in the design arena. The convergence of XR technology and autonomous systems resonates with his passion for design and problem-solving.

Reflecting on past achievements, a significant breakthrough occurred 13 years ago when Ryan and his team addressed a critical issue—design flaws reaching the factory and sometimes customers. This challenge led to the creation of a solution that enabled experiencing designs in a manufacturing context before investing in production. This early venture into virtual prototyping has since evolved and scaled, democratizing CAD models and finding applications in work instruction development and training while helping more than 500 new product designs transition to manufacturing with almost zero assembly issues.

Looking ahead, the industry's interest in remote assistance and Augmented Reality (AR) work instructions is evident. Countering some perceiving AR as gimmicky, the growing demand among end-users indicates its practical applications. The desire for intuitive, hands-free interfaces underscores the need for technology to simplify complex tasks and enhance productivity.

CHAPTER 7   LEARNING FROM EXPERIENCE

As Ryan delves into his passion for helping others succeed, his career revolves around accelerating awareness. This encompasses facilitating knowledge transfer through novel approaches that shorten learning curves from months to seconds. His commitment to helping novices become experts aligns with the broader goal of making information glanceable digestible and tacit expertise unambiguously transferrable.

The concept of the industrial metaverse, though often labelled a buzzword, resonates with the need for heightened awareness and informed decision-making. Leveraging Industry 4.0 principles, the industrial metaverse empowers organizations to harness sensor data and provide autonomous, context-aware insights. The goal is to improve quality, streamline workflows, enhance productivity, and ensure relevant information reaches users at the right time and, importantly, in the right way.

Regarding successful XR and advanced visualization deployment and scaling, Ryan's top recommendations include prioritizing user-centric design, understanding the dynamics of valuation from concept through scaling, and navigating the challenges posed by evolving technologies. Designing solutions that users willingly adopt, coupled with a clear understanding of technology valuation, ensures the technology's effectiveness in solving real business problems.

However, challenges persist in large-scale adoption, primarily related to rapidly evolving technologies and enterprise constraints. As IT organizations and enterprise policies strive to keep current with the rapid pace of change, the need for dedicated teams focused on staying abreast of XR technologies and strategic decision-making becomes paramount for companies seeking seamless integration.

In conclusion, Ryan's journey intertwines design thinking, user experience, adoption theory, technology valuation, and a fervent commitment to solving real business problems with XR and further advanced visualization technologies. From democratizing CAD models to accelerating knowledge transfer and exploring the industrial metaverse,

CHAPTER 7   LEARNING FROM EXPERIENCE

the convergence of these elements propels industries toward a future defined by informed decision-making, profound collaboration, and heightened awareness.

Interview with Paul Davies (Boeing)

Figure 7-3. *Paul Davies*

In the realm of XR technology and Model-Based Systems Engineering (MBSE), Paul Davies (Figure 7-3), a Technical Fellow at Boeing Research and Technology, brings a wealth of expertise rooted in his educational background – bachelor's and master's degrees in electrical engineering with a specialization in digital signal processing. Paul's educational journey served as the gateway to his immersive technology career. His focus involves working on image processing, recognizing, and tracking features to aid companies in visualizing and solving business challenges.

CHAPTER 7 LEARNING FROM EXPERIENCE

Boeing, under Paul's guidance, strategically deploys XR to revolutionize design and factory reviews from a holistic life cycle perspective. This entails engineers sharing 3D virtual experiences, with an emphasis on the design phase, while Boeing endeavors to introduce live 3D models in the assembly phase. Factory layout reviews, exemplified by designing production layouts with tools like the Microsoft HoloLens, are gaining traction. However, the paramount application lies in work instructions for services, maintenance, and customer support departments. This includes developing and enhancing skills for mechanics through immersive technology, a crucial aspect in a complex industry where building airplanes spans months and demands highly specialized expertise.

Paul emphasizes the transformative impact of XR in validating work correctness, envisioning a future where 2D cameras or AR devices with 3D depth sensors verify tasks, auto-populating schedules, and facilitate seamless job sign-offs. Despite challenges, Paul recognizes the pivotal role of user acceptance in adoption, stressing that the benefits must outweigh the perceived barriers. In navigating XR integration with IT infrastructure, he acknowledges existing limitations and the ongoing quest for optimal solutions.

The discussion extends to Model-Based Systems Engineering (MBSE), where Paul acknowledges its adoption at Boeing, particularly in new programs. His work aligns with MBSE principles, leveraging metadata from design models to enhance XR experiences. While acknowledging the XR industry's immaturity, Paul's pragmatic approach involves seizing opportunities when they arise rather than waiting for standardized methods.

The conversation concludes with insights into the challenges of interoperability in XR and the broader concept of an industrial metaverse. Paul views the industrial metaverse as a digital twin of the entire industry, a virtual realm where real-world actions seamlessly transpose. He recognizes the buzzword nature of the "metaverse" and the challenge of bundling diverse technologies under a singular label.

CHAPTER 7 LEARNING FROM EXPERIENCE

In terms of successful XR deployment, Paul emphasizes user engagement, iterative involvement of teams, and the imperative of making the technology a team effort. Looking ahead, he envisions emerging use cases focusing on validating task completion and automating processes through XR.

Paul candidly discusses the hurdles, including the current landscape of companies vying to retain control over end-to-end processes and the anticipated challenges in achieving interoperability.

Interview with Francis Vu (Immersified)

Figure 7-4. *Francis Vu*

Francis Vu ("Frankie", Figure 7-4) is the founder of Immersified, an XR agency that is driving the adoption of virtual and mixed reality by helping businesses with their first experiences of the technology. Having showcased

XR technology to wide audiences since 2015, Frankie is passionate about the impact that immersive tech can bring to a wide range of use cases that are growing every day. Before founding Immersified, Frankie led training programs at Meta (Facebook) to introduce a range of immersive products to varied audiences while also working in the media as an on-camera host.

What role does your company play in the XR growing industry?

Immersified aims to support the transition of businesses into the age of XR by advising companies on the best practices when introducing immersive tech devices and products. Aiming primarily at new and inexperienced users of immersive tech, the impact is achieved through ongoing training programs, instructional guides, and live workshops, where innovators and business leaders gain the foundational skills and inspiration to implement immersive tech into their company workflows.

What interests you personally about XR development? The next big thing?

Personally, I have always been fascinated by technologies that have the potential to impact human lives and change the way that everyday activities are carried out. Over the last several decades, there have been numerous technologies that have drastically changed human patterns of work, socializing, exercising, and leisure time, among others. I believe that XR has the power to create one of these paradigm shifts, and I'm excited to be involved in the industry at a relatively early stage.

I believe the 'next big thing' will be the intersection of generative AI tools with immersive technology and all of the varied ways this can be applied. Although AI has already played an essential and pivotal role in XR spatial tracking systems for a number of years, the rapid advancement of LLMs (Large Language Models) is allowing for advanced real-time interaction within contextual prompt guidelines. This is paving the way for unique immersive, interactive, and authentic educational simulations, as well as new opportunities for healthcare provision, gaming, virtual influencers, and beyond. The convergence of these technologies is extremely exciting as they merge spatial presence with the power of generative AI.

CHAPTER 7   LEARNING FROM EXPERIENCE

What is the industrial metaverse for you, and why should enterprises care?

From my perspective, the industrial metaverse represents the next evolution of the workplace, where immersive technologies are seamlessly integrated into a number of essential processes in design, testing, manufacture, sales, training, and beyond. Enterprises should be taking active steps to partake in the development of the industrial metaverse because of the far-reaching implications that this shift is likely to have on a wide number of industries. Early investment to lay the infrastructure and groundwork for these changes will lead to deeper learning and a greater competitive edge for the enterprises concerned.

In your view, what problems can it solve?

Physical presence.

I believe the industrial metaverse can – to an extent – solve the problem of physical presence and immediacy in an increasingly global landscape. Convincingly simulating physical presence can provide a useful alternative to travel, allowing teams to work together on joint projects that were previously only possible in-person. Such synchronous and in-person sessions are likely to come at greater cost financially as well as environmentally and in time spent. By using immersive solutions, enterprises can benefit from greater return on investment.

In the same vein, immersive tech unlocks new possibilities for instructor-led training sessions and collaborative work. With the rise of distributed teams, XR can also help foster a more authentic feel to benefit team spirit.

Procedural Training

There have already been some great examples of immersive tech helping to improve the effectiveness of training for manual or even surgical procedures across a range of industries. Not only can this reduce training costs while building effectiveness but it can also improve safety via the use of virtual scenarios, therefore not exposing participants or facilitators to potential physical hazards.

Do you have insights on how to successfully deploy XR in industrial/company environments?

Like with any new computing platform, the deployment of XR in company environments should be carefully planned and gradually implemented. In my experience, some general and non-exhaustive advice would include

- Consult experienced XR experts and trainers who can share their experience and set the program up for success. Experts will be able to advise on use cases, implementation, device selection, appropriate applications, cloud device management and phased deployment strategy.

- Appoint immersive tech "champions" within the company who are early adopters and can help evangelize the technology within the context of the company's existing workflows. Champions should receive training in order to build foundational solid practices.

- Obtain support and commitment from leadership to ensure full backing for the program.

- Coordinate and agree on a united deployment strategy across the company so as not to duplicate efforts or cause confusion.

- Employ a dedicated team to manage device inventory and ensure efficient upkeep and overview of hardware.

CHAPTER 7   LEARNING FROM EXPERIENCE

- Pilot the use of these devices on small groups, gaining detailed feedback at every opportunity so as to iterate and improve before scaling to a broader group.
- Build and maintain a library of resources to aid colleagues in their discovery of the tech.
- Stay up-to-date with industry updates to ensure that devices are being used in the most effective way.

What kind of use cases do you think we will see emerging at manufacturers in the near future?

We have already seen the use of immersive tech to aid the improvement of manual processes in the manufacturing pipeline, but I believe that this will gain wider usage in coming years as the tracking and graphical fidelity improves for augmented and mixed reality devices. Further to the use case of improving manual processes, I believe there will also be growth in the use of XR earlier in the manufacturing process, for collaborative design and prototyping.

Thinking about your company's use cases, what is the most significant barrier to large-scale adoption of XR technology?

There are a number of barriers to large-scale adoption of XR technology but I believe the single most significant barrier can be summarized as the amount of **friction experienced when integrating the tech into existing workflows**. There is no denying the number of legitimate and impressive use cases for immersive tech, and the quality of the experience has improved greatly in recent years. However, unless all of these benefits can be experienced almost instantly and with little to no friction, I believe mass adoption will still be out of reach. Consider the devices that are used on a daily basis by the workforce and developed populations at large: desktop computers, laptops, and smartphones. The adoption of all of these devices took decades of research and multiple iterations before reaching critical mass. Although timelines for new product adoption are arguably decreasing, the usage patterns of XR devices (i.e., donning head-mounted devices) are significant enough a departure from established patterns (i.e., 2D monitors, peripherals, and touchscreens) that an almost seamless transition between mediums will be required in order to allow the entry of immersive tech into this tier of ubiquitous use. One could even go so far as to say that **the use of XR devices must present a comparative advantage to more traditional mediums** before it can reach the height of its potential adoption. Here are a number of factors that I believe will help to break down this barrier, including

- Smaller, lighter form factors for XR devices
- Simpler setup processes
- Continued improvement of control interfaces (e.g., hand tracking, gesture control)
- Seamless Integration with existing platforms

How is digital twinning connected to this and what can it provide businesses?

Digital twinning can provide great benefits in a number of capacities. For example, using a digital twin of a well-known office building as a venue to host virtual training sessions, can help to build stronger engagement for employees of a company, leading to higher effectiveness of sessions. In my experience, this approach has been used to great effect. Other strongly beneficial uses of digital twins are for facilities, security and event planning. Having accurate building representations along with the ability to switch between different perspectives can be extremely effective for any logistical planning and training.

CHAPTER 7   LEARNING FROM EXPERIENCE

Interview with Cedric Chane Ching (Aptero)

Figure 7-5. Cedric Chane Ching

Cedric is French–Mauritian and grew up in Mauritius (Figure 7-5). He speaks six languages: French, English, Japanese, Chinese, Creole, and Spanish. He worked as the right hand CEO of Dether, a blockchain startup that raised US$ 13 mil. Cédric is Franco–Mauritian and has worked in four countries. Cédric has technical expertise and entrepreneurial spirit. From the age of 14, when he coded his first video game, he knew he wanted to create boundless tech innovation that serves the world. Since then, Cédric graduated from the best engineering and business institutions and traveled to more than 60 countries. As a multilingual engineer with an international perspective and has a scientific and marketing approach, facilitating global expansion for Aptero.

What role does your company play in the XR growing industry?

Aptero plays an important role in the XR industry, creating innovations in scalability and feature development. Our unique approaches have enabled the creation of numerous advanced features, including virtual machines, within the XR landscape.

What interests you personally about XR development? The next big thing?

CHAPTER 7   LEARNING FROM EXPERIENCE

I'm fascinated by XR development because it breaks down physical and geographical barriers, enabling immersive experiences that can transform how we learn, work, and interact. The next big thing, in my opinion, is the intertwining of XR with AI, creating smart, adaptive, and personalized virtual experiences for users in real time.

Example with one of our clients using our solution for an immersive job description. Now you can go to our XR space, ask the AI agent: what's the best job for me if I like speaking to people? The AI agent directly redirects you to the relevant jobs with an immersive view in 360° of the job.

What is the industrial metaverse for you, and why should enterprises care?

The industrial metaverse to me represents a digital twin of our physical industrial world, where processes, operations, and data converge in an interactive, 3D virtual space. Enterprises should care because it provides an arena where real world and digital elements coalesce, enabling optimized processes, predictive maintenance, and collaborative work environments, transcending geographical limitations.

In your view, what problems can it solve?

XR and the metaverse can solve problems related to remote collaboration, training, and operations management. They provide practical, immersive experiences for training, enable real-time collaboration on a global scale, and facilitate optimized asset management through digital twins, consequently reducing downtime and improving operational efficiency.

Example has been given through the job description, but immersive learning can now be active, and this is what we can provide.

Do you have insights on how to successfully deploy XR in industrial/company environments?

Successful XR deployment depends on aligning technology with specific use cases, ensuring scalability, and focusing on user experience. Begin with pilot projects, prioritize usability, ensure IT infrastructure can handle XR demands, and progressively scale while ensuring alignment with organizational goals and user needs.

We always focus on accessibility within our POCs, so that every user can live the experience.

What kind of use cases do you think we will see emerging at manufacturers in the near future?

Anticipating the evolution of XR in manufacturing, I envision use cases such as enhanced remote assistance, wherein experts can guide on-site technicians in real time. Also, immersive training environments, where personnel can safely simulate and navigate through varied scenarios, and virtual prototyping to visualize and test designs before physical production.

Thinking about your company's use cases, what is the greatest barrier to large-scale adoption of XR technology?

The greatest barrier, in my perspective, is the integration complexity, with technological compatibility, data security, and achieving uniformity in user experience, particularly when blending XR with existing systems and processes. Overcoming this involves strategic planning, phased implementation, and continuous optimization.

Anything else you would like to add?

While XR does have multiple challenges, it's imperative to recognize that the XR and metaverse journey isn't solely about technology. It's about crafting experiences, creating connections, and enabling opportunities in a parallel digital universe, shaping our future interactions and operations in both realms.

CHAPTER 7   LEARNING FROM EXPERIENCE

Interview with Alexandre Mao (Torrus VR)

Figure 7-6. *Alexandre Mao*

Alexandre Mao (Figure 7-6) is CEO and co-founder of Torrus VR, a company specializing in developing immersive experiences using AR, VR, VideoMapping, and IoT.

He started his journey in the VR industry back in 2015, developing a motion tracking system for large-scale, multiplayer experiences.

What role does your company play in the XR growing industry?

We are specialized in developing interactive multiplayer experiences in VR and AR. We developed a motion tracking system for companies to create immersive and interactive large-scale experiences, where we are able to freely move. Our goal back then was to make it easy to create hyper-immersive experiences; then we switched our services when the VR standalone headset came out, providing services to create multiplayer colocated experiences.

We currently serve multiple companies providing our experiences creating custom solutions using XR for them in various fields.

What interests you about XR development?

XR is the evolution of what we had in 2D with computers, it enables us to enter into the 3D world and interact with it in a whole different way.

CHAPTER 7   LEARNING FROM EXPERIENCE

Here are some points that I find amazing in VR:

- Immersive Storytelling . XR allows us to create deeply immersive experiences that engage users in ways traditional media cannot. The ability to transport users to entirely different worlds or scenarios is incredibly powerful.

- Enhanced Interaction. XR offers new ways for users to interact with digital content, making interactions more natural and intuitive. Gameplays are totally different than the other devices

What is the industrial metaverse, and why should enterprises care?

The metaverse can be defined simply by persistent multiuser virtual world. The industrial metaverse can be more specifically defined as a digital twin of the physical industrial world, where immersive, interactive, and real-time simulations of industrial operations, processes, and systems are created and managed. It can enable enterprises the following benefits:

- Real-time Monitoring and Optimization: By using digital twins and IoT, enterprises can monitor their operations in real time and optimize processes, leading to increased efficiency and reduced downtime. Different scenarios can be tested without impacting the ongoing process of the company.

- Predictive Maintenance: AI-driven analytics can predict equipment failures before they occur, allowing for proactive maintenance and minimizing unplanned outages. Simulations will provide more details on what can happen and give instructions to prevent possible failures.

CHAPTER 7   LEARNING FROM EXPERIENCE

- Simulation and Testing: Enterprises can simulate and test new processes, systems, or products in the metaverse before implementing them in the real world, reducing the risk of costly errors. Digital twins in this case are really important so that the simulation is accurate.

- Immersive Training: XR technologies provide realistic training environments for employees, improving skill acquisition and safety without the need for physical resources.

- Rapid Prototyping: Digital twins and simulations allow for rapid prototyping and iteration of new products or processes, accelerating innovation.

What problems can it solve?

The industrial metaverse can solve multiple problems:

- Training employees on complex machinery or dangerous processes

- Traditional prototyping and testing processes that are time-consuming and expensive

- Inefficient processes lead to increased operational costs and lower productivity.

When will the adoption of the industrial metaverse become make-or-break for businesses?

The adoption of the industrial metaverse becoming a "make or break" factor for businesses will likely depend on several factors, including technological advancements (computing power), industry-specific needs, and competitive pressures.

Easy access is a key part, with possibilities like cloud computing to make realistic experiences run smoothly on standard devices

CHAPTER 7 LEARNING FROM EXPERIENCE

Some important steps that are needed in my point of view are

- Technological Maturity, Advanced and reliable XR, IoT, AI, and cloud computing technologies become widely accessible and affordable to implement industrial metaverse solutions effectively.

- Standardization and Interoperability. Industry standards and protocols for interoperability between different metaverse platforms (openXR) and technologies are established.

What is the main limitation to scaling up XR deployment in the industry?

The main limitation to scaling up XR deployment in the industry is how smoothly it will be integrated in the employee journey, how easy and natural it will be for them to use it.

Current XR devices can be bulky, uncomfortable for long-term use, and limited by battery life and performance. Once the user experience is as smooth for people as using a computer, the adoption of XR will be way easier.

Do you have insights on how to successfully deploy XR in industrial environments?

Successfully deploying XR in industrial environments requires a strategic approach that addresses the unique challenges and leverages the opportunities of the technology. Here are some insights on how to achieve this:

- Define clear objectives and use cases where XR can add significant value; technology is a tool to improve our efficiency in our work, not an end goal.

- Clear objectives will help to have measurable goals so that the impact can be measured, for example, in terms of productivity gains, cost savings, safety improvements, or other metrics relevant to the industry.

CHAPTER 7   LEARNING FROM EXPERIENCE

- Create a great User Experience, design XR applications with the end user in mind, ensuring they are intuitive, easy to use, and enhance the user's workflow.

What kind of use cases do you think we'll see emerging at manufacturers in the near future?

Some use cases that will see emerging are

- Training, using VR to create realistic training scenarios for workers to practice operating machinery, handling hazardous materials, or responding to emergency situations without real-world risks.

- Remote Assistance, AR and MR can enable remote experts to guide on-site technicians through complex repairs by overlaying instructions and annotations on the technician's view of the equipment.

- Design and prototyping, using VR to create and manipulate 3D models of new products, enabling faster iterations and more collaborative design processes without the need for physical prototypes.

Thinking about your company use cases, what is the greatest barrier to large-scale adoption of XR technology?

The greatest barrier to large-scale adoption of XR technology is the high initial costs, simply talking of the cost of custom development of XR applications that can be expensive and the training related to the usage of a new technology. Training programs need to be developed and implemented to educate employees on how to use the new XR applications ensuring that the technology is utilized effectively and safely.

APPENDIX A

Artificial intelligence in the Metaverse

This appendix discusses the integration of Artificial Intelligence (AI) and Extended Reality (XR) in creating immersive virtual worlds. It highlights the potential of combining AI's smart capabilities with XR's advanced visualization techniques to develop a metaverse, a collective virtual shared space. The appendix emphasizes the importance of understanding the relationship between AI and XR to shape future technologies and explores the transformative impact of this integration on various industries and societal interactions.

Key points include

- AI and XR can create immersive experiences that transcend traditional boundaries.
- Generative AI algorithms and advanced visualization techniques can elevate XR experiences.
- AI plays a crucial role in managing the infrastructure of the metaverse, enhancing generative interactions, enabling computer vision capabilities, and optimizing data management processes.

- AI-driven content generation, such as using Generative Adversarial Networks (GANs), can create realistic textures, landscapes, and objects within augmented and virtual reality environments.

- AI technology can improve user interaction with software through gesture recognition systems and computer vision algorithms.

Overall, the appendix underscores the immense potential of AI and XR in shaping the future of digital realms and enhancing user experiences.

In the realm of technological advancements, AI has emerged as a pervasive force, revolutionizing various aspects of our lives, including the way we work. Consequently, it would be shortsighted to overlook the interconnectedness of AI and XR, specifically considering its potential integration in the creation of immersive virtual worlds. Indeed, it would be unwise to disregard the interdependence of AI and XR, particularly when considering their potential collaboration in developing captivating virtual realms. By combining the capabilities of AI and XR, we can unlock new possibilities for creating immersive experiences that go beyond traditional boundaries (Figure A-1).

APPENDIX A   ARTIFICIAL INTELLIGENCE IN THE METAVERSE

Figure A-1. *Abstract illustration of the AI-driven metaverse*

The concept of XR involves simulations and modeling that blur the boundaries between physical and digital realms. By harnessing generative AI algorithms and leveraging advanced visualization techniques, XR experiences can be elevated to new heights. As such, AI's smart capabilities enable us to explore the possibility of creating a metaverse – a collective virtual shared space – driven by machine learning algorithms.

Through this convergence of XR and AI, we open doors to unprecedented experiences and opportunities for innovation. By further exploring this connection, we can unlock immense potential in developing immersive environments that transcend traditional limitations. As we navigate this intriguing landscape, it becomes crucial to carefully examine the transformative impact that integrating these technologies may have on various industries and societal interactions. As we delve into this captivating realm, it is essential to thoroughly analyze the profound influence that the integration of these technologies can potentially have on diverse industries and societal interactions.

In conclusion, understanding the relationship between AI and XR is vital in shaping future technologies. Advances in generative AI, coupled with sophisticated visualization techniques, hold great promise in realizing the vision of a metaverse – an interconnected digital realm – exceeding our wildest imaginations.

The Role of AI in the Metaverse

In the metaverse, AI technology plays a crucial role in bringing to life the promises that investors and developers have envisioned. There are several key areas where AI will significantly contribute to making the metaverse a reality.

Firstly, leveraging AIOps (Artificial Intelligence for IT Operations) is essential in managing the underlying infrastructure of the metaverse. With vast amounts of data, complex networking systems, and various interconnected technologies, AI can help monitor, analyze, and optimize the performance of these systems. This ensures that the metaverse remains stable and accessible for users.

Another important aspect where AI comes into play is in generative interactions within the metaverse. Through natural language processing and machine learning algorithms, AI enables virtual characters or NPCs (non-player characters) to engage with users in realistic and dynamic ways. This enhances immersion and makes interactions within the metaverse more engaging and meaningful.

Computer vision is yet another area where AI shines in the context of the metaverse. By leveraging computer vision algorithms, AI can recognize objects, gestures, facial expressions, and even emotions of users' avatars or real-life representations. This enables more intuitive user experiences within virtual environments and facilitates seamless integration between real-world actions and virtual responses.

Furthermore, AI technology plays a critical role in indexing and managing storage as well as transmission of vast amounts of data within the metaverse. As users generate enormous amounts of content ranging from images to videos to user-generated virtual environments, AI-powered algorithms can help categorize, organize, retrieve, and distribute this data efficiently.

Overall, AI's integration into various aspects of the metaverse helps create immersive experiences while ensuring that infrastructure is manageable at scale. From enhancing generative interactions to enabling computer vision capabilities to optimizing data management processes, artificial intelligence holds tremendous potential in shaping the future of this digital realm.

How AI Helps with Metaverse Content Creation

In the evolving landscape of the metaverse, artificial intelligence (AI) technologies play a crucial role in enhancing content creation. One prominent AI-driven technology is generative AI, which utilizes Generative Adversarial Networks (GANs) to create highly realistic textures, landscapes, and objects within augmented reality (AR) and virtual reality (VR) environments.

The metaverse, often referred to as a collective virtual shared space, requires a robust infrastructure capable of handling data distribution and ensuring digital continuity. AI can contribute to this by generating detailed content that closely mimics real-world elements, thereby enriching the immersive experience for users. Through GANs' capability to generate high-quality and lifelike digital assets, content creators in the metaverse can leverage these AI technologies to create visually stunning worlds. By employing generative AI algorithms, developers can save time and effort required to manually design intricate details within the metaverse environment (Figure A-2).

APPENDIX A   ARTIFICIAL INTELLIGENCE IN THE METAVERSE

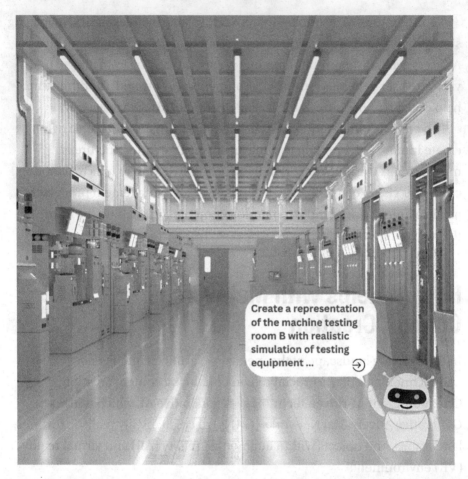

Figure A-2. *How AI could create virtual simulations on user request*

Furthermore, AI-driven content generation facilitates digital continuity within the metaverse by ensuring consistent quality across different virtual experiences. This means that regardless of location or device used by users to access the metaverse, they can expect visually compelling and cohesive digital content.

In conclusion, the integration of artificial intelligence technology such as generative AI into the infrastructure of the metaverse enhances content creation by providing detailed textures, landscapes, and objects that closely resemble real-world elements. This assists in maintaining digital continuity while offering immersive experiences for users exploring these virtual spaces.

The Superpowers AI Can Provide to XR Interaction

With AI technology, users can now effortlessly switch between different tools by performing various gestures during a task. This eliminates the need to pause work and search through menus or press buttons on a controller or keyboard. In a recent study, researchers developed a neural network gesture recognition system that can identify gestures by making predictions based on an incoming stream of hand joint data. This innovative approach enables users to interact with software using natural hand movements rather than relying solely on memorized hotkeys. By leveraging this gesture recognition system, users can seamlessly transition between different tools without interrupting their workflow. This not only enhances productivity but also provides a more intuitive and efficient user experience. It's worth noting that this technology is still in its early stages of development and may not be widely available yet in all software applications. However, it holds promising potential for improving user interaction with desktop software in the future.

Computer Vision for AR

Computer vision algorithms are indeed crucial for supporting the metaverse, especially its augmented reality (AR) component. As many interactions within the metaverse will be hybrid experiences rather than full immersion, the hardware enabling such immersion needs to be aware of our position and capable of understanding what we see in order to provide contextual information. Hololens, in its two incarnations at the time of writing, and the short-lived experiment with Google Glass have demonstrated to both the public and industrial world the potential of AR applications. Indeed, the Microsoft HoloLens and Google Glass have showcased the immense potential of augmented reality (AR) applications to both the general public and various industries. These devices have paved the way for a new era of interactive digital experiences. The HoloLens, available in two versions at present, is a self-contained wearable computer that overlays holographic content onto the real world. It allows users to interact with virtual objects and information seamlessly integrated into their physical surroundings. This technology has been embraced by various sectors such as architecture, healthcare, education, and entertainment, offering innovative solutions for design visualization, surgical planning, immersive learning experiences, and more. On the other hand, Google Glass was an early attempt at bringing AR capabilities to everyday eyewear. Although its initial iteration faced challenges in terms of adoption due to privacy concerns and limited functionality for everyday use cases, it did manage to demonstrate some interesting possibilities. For instance, users could access information hands-free or capture photos or videos using voice commands. Overall, both HoloLens and Google Glass have played a significant role in raising awareness about AR technology's potential across industries. Their experiments have fueled further research and development in this field while inspiring new applications that can enhance productivity, efficiency, and creativity while transforming our daily lives.

It is worth noting that around 90% of the usefulness of these applications stems directly from pattern recognition and image classification algorithms embedded within them. These algorithms enable the software to analyze and interpret visual data, helping to identify patterns and classify images accurately. By leveraging advanced machine learning techniques, such as convolutional neural networks (CNNs), these algorithms can achieve remarkable levels of accuracy and efficiency. As a result, users can benefit from various applications that rely on image analysis, ranging from facial recognition technology to object detection systems. These algorithms allow for recognizing objects or patterns in real time and providing relevant information or augmentations based on that recognition. In conclusion, computer vision algorithms play a vital role in enhancing AR experiences within the metaverse by enabling hardware to understand visual input and provide contextual information accordingly.

APPENDIX B

Sources

Chapter 1

- Ffiske, Tom. The Metaverse: A Professional Guide: An expert's guide to virtual reality (VR), augmented reality (AR), and immersive technologies (p. 60). Kindle Edition.

- Metaverse: origins and definition of a modern phenomenon. By Léa Paule. https://blog.laval-virtual.com/en/origins-and-definition-of-the-metaverse/

- A narrative review of immersive virtual reality's ergonomics and risks at the workplace: cybersickness, visual fatigue, muscular fatigue, acute stress, and mental overload. Alexis D. Souchet, Domitile Lourdeaux, Alain Pagani, Lisa Rebenitsch.

Chapter 2

- R. Y. Rubinstein. "Simulation and the Monte Carlo Method," Wiley & Sons (1981).

APPENDIX B SOURCES

- Survey of Model-Based Systems Engineering (MBSE) Methodologies. Jeff A. Estefan. Jet Propulsion Laboratory California Institute of Technology Pasadena, California, U.S.A.

- A.J. Ramirez, A.C. Jensen, B.H.C. Cheng. A Taxonomy of Uncertainty for Dynamically Adaptive Systems, SEAMS, 2012.

- Martin A. Bauer. Tracking Errors in Augmented Reality, Technische Universit at Munchen, Institut fur Informatik Chair for Computer-Aided Medical Procedures & Augmented Reality.

- IEEE Standard 1061-1992 Standard for a Software Quality Metrics Methodology. Institute of Extended Reality in the system life cycle.

Chapter 3

- Dieter Schmalstieg, Tobias Hollerer. "Augmented Reality: Principles and Practice," Addison-Wesley Professional (2016).

- John Vince, "Virtual Reality: Principles and Practice," Addison-Wesley Pub. Co (1995).

Chapter 4

- A. Migliaccio, G. Iannone (2023). Systems Engineering Neural Networks, Wiley.

- J. Hold, S. Perry (2020). Don't Panic, The Absolute Beinner's Guide to Model-Based Systems Engineering, INCOSEUK.

- M. Luffi, R. Valerdi (2020). Virtual Reality in Model Based Systems Engineering: A Review Paper – Part of the Communications in Computer and Information Science book series (CCIS, volume 1294).

- M. Sesana, A. Mousssa (2019). Collaborative Augmented worker and Artificial Intelligence in Zero defect Manufacturing environment, MATEC Web of Conferences 304, 04003.

- Z. Hu, R. Arista, and others (2022). Ontology-based system to support industrial system design for aircraft Assembly. IFAC.

- (PDF) Engineering Design Process for virtual reality headset. Available from: https://www.researchgate.net/publication/347838152_Engineering_Design_Process_for_virtual_reality_headset [accessed Jun 11 2023].

Chapter 5

- ROI calculator and other material on the use cases provided by AREA, https://thearea.org/

Appendix A

- Zhaomou Song, John J Dudley, and Per Ola Kristensson. "HotGestures: Complementing Command Selection and Use with Delimiter-Free Gesture-Based Shortcuts in Virtual Reality." IEEE Transactions on Visualization and Computer Graphics (2023). DOI: 10.1109/TVCG.2023.3320257.

Glossary

Artificial Intelligence

Artificial intelligence, often referred to as AI, is a branch of computer science that focuses on creating intelligent machines that can simulate human behavior and cognitive processes. These machines are designed to perform tasks that typically require human intelligence, such as learning, problem-solving, perception, and decision-making.

Augmented Reality

Augmented Reality refers to any technology that "augments" the user's visual (and in some cases auditory) perception of their environment. Typically, digital information is superimposed over a natural, existing environment. Information is tailored to the user's physical position as well as the context of the task, thereby helping the user to solve the problem and complete the task.

Centralize Data Lake

A centralized data lake is a single, unified repository that stores large amounts of raw data in its native format until it is needed. This approach allows organizations to store structured and unstructured data from various sources in one location for easy access and analysis. By centralizing data in a data lake, organizations can break down silos between different departments or systems, enabling better collaboration and insights across the organization. Data lakes are designed to handle massive volumes of data efficiently, making them ideal for storing diverse types of information such as text, images, videos, and more. In essence, a centralized data lake serves as a scalable and flexible storage solution that empowers businesses to harness the full potential of their data assets for analytics, machine learning, and other advanced applications.

Cross Reality

Cross reality, also known as XR, is an umbrella term that encompasses all forms of immersive technologies, including virtual reality (VR), augmented reality (AR), and mixed reality (MR). These technologies blend the physical and digital worlds to create interactive experiences that can range from fully immersive virtual environments to overlays of digital information in the real world.

XR allows users to engage with content in a more interactive and engaging way, breaking down barriers between the physical and digital realms. By leveraging XR technologies, users can explore new worlds, visualize data in innovative ways, and interact with digital content in a more intuitive manner.

In essence, cross-reality represents the convergence of physical and digital realities, offering endless possibilities for entertainment, education, training, and various other applications. As XR continues to evolve and become more accessible, its impact on industries such as gaming, healthcare, education, and marketing is expected to grow significantly.

Cybersecurity

Cybersecurity is a critical aspect of our digital world, encompassing the protection of computer systems, networks, and data from theft, damage, or unauthorized access. It involves implementing measures to ensure the confidentiality, integrity, and availability of information in the face of cyber threats. In simple terms, cybersecurity refers to the practice of safeguarding digital information from cyberattacks.

These attacks can come in various forms, such as malware, phishing scams, ransomware, or hacking attempts. The goal of cybersecurity is to prevent these attacks and protect sensitive data from falling into the wrong hands.

Cybersickness

Cybersickness, also known as digital motion sickness, is a condition that can occur when an individual is exposed to virtual environments or digital screens for an extended period. Symptoms of cybersickness may include nausea, dizziness, headaches, and eyestrain.

Digital Continuity

Digital continuity refers to the seamless and uninterrupted flow of digital information and services across various platforms, systems, and technologies. It ensures that data remains accessible, usable, and secure throughout its life cycle. In today's fast-paced digital world, businesses rely heavily on digital continuity to maintain operations, deliver services efficiently, and meet customer expectations.

It involves strategies and technologies that enable organizations to manage data effectively, prevent disruptions in service delivery, and adapt to changing technological landscapes. By implementing robust digital continuity practices, businesses can safeguard their critical information assets, enhance operational resilience, and stay competitive in the ever-evolving digital marketplace. It is essential for organizations to prioritize digital continuity as part of their overall business strategy to ensure long-term success in the digital age.

Field of View

Field of View is the term used to explain how large an AR image is when viewed, which explains how "big" an augmented reality image is when viewed through a display device. A smaller field of view can limit the size and quality of an AR image. Generally speaking, a larger field of view offers more flexibility in locating the AR image and any accompanying information for optimum viewing for a particular application.

Functional Requirements

Functional requirements are specifications that define the functions and capabilities a system, software, or product must have to meet the needs of its users. These requirements outline what the system should do, rather than how it should do it. They serve as a roadmap for developers and designers to ensure that the end product meets the desired objectives.

GLOSSARY

In software development, functional requirements typically include features such as user authentication, data management, reporting capabilities, and integration with other systems. These requirements are essential for guiding the development process and ensuring that the final product meets user expectations.

Immersive Technology

Immersive technologies refer to technologies that aim to create a sense of immersion or presence in a digital environment. These technologies typically include virtual reality (VR), augmented reality (AR), and mixed reality (MR).

Intrapreneurship

Intrapreneurship is a concept that refers to employees within a company who take on the mindset and responsibilities of an entrepreneur. These individuals are innovative, proactive, and driven to create new ideas, products, or processes within the existing organizational structure.

Intrapreneurs are often encouraged to think creatively and take risks in order to drive growth and innovation within their company. They may identify new opportunities for business development, propose solutions to challenges, or spearhead new projects that can benefit the organization as a whole.

By fostering a culture of intrapreneurship, companies can tap into the creative potential of their employees and stay ahead of the competition in today's rapidly changing business landscape. This approach not only benefits the company by driving innovation and growth but also empowers employees to take ownership of their work and contribute meaningfully to the organization's success.

MBSE

MBSE, or Model-Based Systems Engineering, is an approach to system design and development that focuses on creating and using models as a central tool to support system requirements, design, analysis, verification, and validation activities throughout the life cycle of the system. In MBSE, models are used to capture and communicate system requirements in a structured way that allows for better understanding and management of complex systems. These models can range from simple diagrams to detailed simulations depending on the needs of the project. By using MBSE, organizations can improve collaboration among different teams working on a project by providing a common language and framework for discussing system requirements and design decisions. This approach helps reduce errors, improve efficiency, and ensure that all stakeholders have a clear understanding of the system being developed. Overall, MBSE offers a systematic way to develop systems by leveraging modeling techniques to ensure that all aspects of the system are well-defined and integrated from the early stages of development through implementation and maintenance.

Metaverse

The Metaverse is a concept that has been gaining significant attention in recent times. It refers to a collective virtual shared space created by the convergence of physical and digital realities. In this immersive digital universe, users can interact with each other and digital objects in real time.

Mixed Reality

Mixed Reality is a form of AR in which digital overlays are blended into the environment and have a spatial relationship with the surrounding objects. While the user moves, the overlays remain located in the same position attached to the real environment. Virtual and real objects are potentially indistinguishable.

Monte Carlo Method

The Monte Carlo method is a computational algorithm that relies on random sampling to obtain numerical results. It is named after the famous casino in Monaco, known for its games of chance and randomness. In essence, the Monte Carlo method involves using random numbers to solve problems that might be deterministic in principle. This method is particularly useful when dealing with complex systems or processes where traditional mathematical approaches may be impractical. By simulating a large number of random samples, the Monte Carlo method can provide estimates and solutions to problems that would otherwise be difficult to solve analytically. In summary, the Monte Carlo method is a powerful tool in the field of computational mathematics and statistics, offering a versatile approach to solving a wide range of problems through random sampling techniques.

Simulation

Simulation is the imitation or representation of one system's operation or characteristics by another. It involves creating a model that behaves similarly to the real-world system under certain conditions. This allows for experimentation, testing, and analysis without directly interacting with the actual system.

In various fields such as engineering, healthcare, and training, simulations are used to replicate real-world scenarios for research, training, or decision-making purposes. By simulating different scenarios and outcomes, professionals can gain valuable insights and make informed decisions. Overall, simulation plays a crucial role in enhancing understanding, predicting outcomes, and improving performance in a wide range of industries.

Single Source of Truth

Single source of truth (SSOT) is a concept that refers to the practice of structuring information in a way that ensures data integrity and consistency across an organization. In simpler terms, it means having one centralized location where all relevant data is stored and managed, serving as the authoritative source for that particular information.

Having a single source of truth eliminates the risk of conflicting or outdated data being used across different departments or systems within a company. This approach helps improve decision-making processes, enhances collaboration among teams, and reduces errors caused by discrepancies in data.

Software Uncertainty

Software uncertainty refers to the unpredictability and variability that can arise during the development, deployment, and maintenance of software systems. It encompasses factors such as changing requirements, technological advancements, market dynamics, and human factors that can introduce risks and challenges in software projects.

In the realm of software development, uncertainty is inherent due to the complexity of modern systems, evolving technologies, and dynamic user needs. Developers often face challenges in accurately predicting project timelines, budgets, and outcomes due to these uncertainties.

Systems Engineering

Systems engineering is a multidisciplinary approach to designing and managing complex systems over their life cycles. It involves considering all aspects of a system, including hardware, software, processes, people, and the environment in which it operates.

In simple terms, systems engineering focuses on ensuring that all components of a system work together efficiently to achieve the desired outcomes. It helps in identifying and solving problems early in the development process, leading to more reliable and cost-effective solutions. By applying systems engineering principles, organizations can streamline processes, improve communication among team members, and ultimately deliver high-quality products or services to customers. This holistic approach is essential for tackling today's increasingly complex technological challenges across various industries.

System Life Cycle

The system life cycle refers to the process of developing, implementing, and maintaining a system from its inception to its retirement. It involves various stages, such as planning, design, development, testing, implementation, and maintenance.

During the planning stage of the system life cycle, the requirements and objectives of the system are identified. This is followed by the design phase, in which the architecture and functionality of the system are defined.

GLOSSARY

The development stage involves building and coding the system based on the design specifications. Testing is a crucial phase where the system is evaluated for performance, reliability, and security. Once testing is complete and any issues are resolved, the system is implemented or deployed for actual use. The maintenance phase involves ongoing support, updates, and enhancements to ensure that the system continues to meet user needs.

Verification and Validation

Verification and validation are two crucial processes in ensuring the quality and reliability of software systems. Verification refers to the process of evaluating a system to determine whether it meets the specified requirements. It involves checking that the software is being built right, meaning that it conforms to its design specifications and correctly implements the intended functionality. Validation, on the other hand, is the process of evaluating a system to ensure that it meets user expectations and needs. It involves checking that the right software is being built, meaning that it satisfies user requirements and can be used effectively in its intended environment.

In essence, verification confirms that the software is being developed correctly, while validation confirms that it is being developed to meet user needs. Both processes are essential for delivering high-quality software products that meet customer expectations and perform reliably in real-world scenarios.

Virtual Reality

Virtual reality (VR) refers to a computer-generated simulation of an environment that can be interacted with in a seemingly real or physical way by a person using special electronic equipment, such as a helmet with a screen inside or gloves fitted with sensors.

In VR, users are immersed in an artificial environment where they can experience and interact with surroundings that feel as if they are real. This technology has applications across various industries, including gaming, education, healthcare, and training simulations.

Index

A

Abstract illustration, 183
Accessibility, 109
Adoption, 200
Aerospace, 96
AI, *see* Artificial intelligence (AI)
AIOps, *see* Artificial Intelligence for IT Operations (AIOps)
AIShed survey, 133–138
AR, *see* Augmented reality (AR)
Architecture, design, 88
Artificial intelligence (AI), 146, 229
    content creation, 219–221
    convergence, 217
    industries and societal interactions, 215
    in metaverse, 15, 16, 218, 219
    superpowers, 221
    systems engineer, 17–19
    and XR, 216
    XR technology issues, 15, 16
Artificial Intelligence for IT Operations (AIOps), 218
Assembly tasks, 194
Augmented reality (AR), 1, 7, 9, 19, 61, 96, 98, 127, 177, 196, 219, 229
    advantages, 150
    computer vision, 222–224
    industrial applications, 192
    mobile device, 74
    navigation, 145
    technology, 73
Automatic simulations, 100, 101

B

BDD, *see* Block Definition Diagram (BDD)
Beginning of Life (BOL), 87
Bitcoin, 4, 5, 19
Blockchain, 206
Blockchain-based time synchronization protocol, 169
Block Definition Diagram (BDD), 66
BOL, *see* Beginning of Life (BOL)

C

CAD, *see* Computer-aided design (CAD)
CDRs, *see* Collaborative Design Reviews (CDRs)

INDEX

Centralized data lake, 229
Change champion, 164
Change Champion Network, 163, 164
Ching, Cedric Chane, 206–209
CNNs, *see* Convolutional neural networks (CNNs)
Cognitive load, 189
Collaboration, 142–144
Collaborative decision-making, 112
Collaborative design, 105
Collaborative Design Reviews (CDRs), 101, 121
Communication, 86
Communication channels, 103
Compatibility, 117
Computer-aided design (CAD), 62, 64
Computer vision algorithms, 221–223
Configuration, 105–107, 122
CONOPS (concepts of operations), 83
Contextually-triggered XR, 139
Convergence
  artificial intelligence (AI), 217
  narrative designer, 92
  primer, 82–85
  XR-MBSE, 92–95
Convolutional neural networks (CNNs), 222
COVID-19 pandemic, 125
Cross reality, 230

Cutting-edge technology, 112
Cybersecurity, 181–185, 230
Cybersickness, 177–179, 230
Cyber threats, 185

D

Data privacy, 116
Davies, Paul, 198–200
Deep modeling, 94
Deployment of XR industrial metaverse, 156
Design flaws, 196
Digital continuity, 76, 77, 80, 85–87, 94, 96, 100–102, 106, 109, 119, 121–123, 161, 168, 231
Digital engineer, 90
Digital engineering, 88
Digital map, 147
Digital mock-ups, 62
Digital models, 194
Digital realms, 216
Digital twins, 95–97, 158
DIY XR, 4

E

End of Life (EOL), 87
Engineering, 6
Enterprise resource planning (ERP), 88
EOL, *see* End of Life (EOL)
Ergonomics, 134

238

ERP, *see* Enterprise resource planning (ERP)
Extended reality (XR), 2–4, 7, 8, 61, 98, 115, 121, 215
  challenges and best practices, 194
  deployment, 167
  gadgetries, 11–13
  implementation, 75
  life cycle, 62, 63
  model-based approach, 168
  planning and execution, 167
  See also XR technology
Eye-tracking data, 184

F

Failure Mode and Effect Analysis (FMEA), 48
Failure model, 48, 49
Fault trees, 48, 49
Field of View, 231
FMEA, *see* Failure Mode and Effect Analysis (FMEA)
Functional requirements, 231, 232

G

GANs, *see* Generative Adversarial Networks (GANs)
Generative Adversarial Networks (GANs), 216, 219
Gesture-based controls, 152
Gesture recognition system, 221

Google Glass, 222
Group beta
  design reviews, 101
  digital continuity, 101
  validation, 101–104
Group delta
  collaborative XR, 112–114
  company's operations, 110
  proposed metaverse, 112–114
  remote assistance, 110–112
  XR training, 109
Group gamma
  abstract illustration, 108
  software developers, 107
  SystML, 108, 109

H

Head-Mounted Display (HMD), 11
HMD, *see* Head-Mounted Display (HMD)
Horizon Workrooms, 135
However-reality, 8
Human capabilities, 195

I, J

IBD, *see* Internal Block Definition Diagram (IBD)
Immersion, 161
Immersive storytelling, 210
Immersive technologies, 232
Inclusion, 161
Industrial innovation, 195

INDEX

Industrial metaverse, 4–6, 23, 149, 155, 156, 173, 175, 193
   concept, 160
   productivity and profitability, 157
   tools, 195
   XR and MBSE, 195
Industrial multiverse, 11
Industry 4.0, 3
Information kidding concept, 191
Innovation, 135
In-service system, 44
Integration, 161, 189
Interactivity, 160
Internal Block Definition Diagram (IBD), 66
Internet of Things (IoT), 181
Interoperability, 160
Intrapreneurship, 165–167, 232
IoT, *see* Internet of Things (IoT)

K

Key enabler, 85
Kinesthetic learning, 133

L

Languages, 54–57
Laughlin, Brian, 190
Lean Six Sigma (L6S), 97
Leveraging methodologies, 37
Life cycle, 88, 93, 108, 118, 119, 122
   management, 93
   phase, 66
L6S, *see* Lean Six Sigma (L6S)

M

Machine learning (ML), 98
Manufacturing execution systems (MES), 88
Mao, Alexandre, 209–213
Marker-based systems, 12
MBSE4XR
   design research, 115
   metaverse, 115, 116
   societal acceptance, 115
   system of interest, 115–117
MES, *see* Manufacturing execution systems (MES)
Metaverse, 6–15, 86, 168, 169, 233
   artificial intelligence (AI), 218, 219
   and bitcoin, 19
   business applications, 15
   concept, 1
   digital realm, 9
   evolution, 160
   industrial, 23
   learning, immersive, 13
   multiple dimensions, 9
   positioning, 151
   representation, 159
   system of interest, 115, 116
Microsoft HoloLens, 191, 199, 222
Middle of Life (MOL), 87

Mixed reality (MR), 7, 61, 99, 127, 233
ML, *see* Machine learning (ML)
Model-based engineering, 88
Model-based system engineering (MBSE) approach, 2, 23, 24, 56, 60, 64, 81, 83, 108, 190, 193, 196, 232, 233
- advantages, 31–33, 42
- architecture models, 28
- customers, 163
- definition, 27
- global/local, 89
- in industrial applications, 189
- industrial metaverse, 41
- MBSE-XR champion, 90, 91
- model-centric systems, 162
- observability, 25
- simulations, 163
- system model, 25
- and System of System (SoS), 26
- techniques, 41
- XR expertise, 90
- XR/immersive producer, 90

MOL, *see* Middle of Life (MOL)
Monte Carlo method, 233
Monte Carlo simulation method, 36–40
MR, *see* Mixed reality (MR)
Multiverse, 9–11

N

Narrative designer, 92

Navigation, 145–148
Neural interfaces, 194
Non-player characters (NPCs), 218
NPCs, *see* Non-player characters (NPCs)

O

Observability, MBSE, 25
On-screen displays, 140
Ontology, 6–15

P

Physical aircraft, 192
Pico4, 134
PIM, *see* Product information management (PIM)
PLM, *see* Product life cycle management (PLM)
Primer, 82–85
POC, *see* Proof of Concept (POC)
Procedural training, 202–205
Product information management (PIM), 87
Product life cycle management (PLM), 87, 93, 96, 102, 119–121
Project Blackwell, 191
Proof of Concept (POC), 3, 127, 128

Q

Quest Pro, 134

INDEX

R

Realistic experiences, 151, 152
Real-time instructions, 130
Real-world scenarios, 25, 40, 54, 57
Reinvention, 176
Remote assistance, 110–112, 129, 130
Remote troubleshooting, 45
Reproducibility, 35
Return on investment (ROI), 171, 172
ROI, *see* Return on investment (ROI)
Room-scale XR, 12

S

Scalability, 116, 194
6DoF, *see* Six degrees of freedom (6DoF)
Self-teaching, 3
Simulation, 22–24, 29–31, 234
    autopilot failure, 49
    bird/lightning strike, 49
    cycles estimation, 52
    error estimation, 53
    hydraulic system failure, 49
    results, 53
    scenarios, 54
    statistical parameters, 50
    trial, 51
Single source of truth (SSOT), 234

Six degrees of freedom (6DoF), 12
Skill acquisition, 193
SMEs, *see* Subject matter experts (SMEs)
Software uncertainty, 33–36, 235
Solid modeling, 92
SoS, *see* System of systems (SoS)
SSOT, *see* Single source of truth (SSOT)
Stand-alone XR devices, 13, 175
Standardization, 11
Subject matter experts (SMEs), 41
Superpowers AI, 221
Synchronization, 160
SysML, *see* Systems Modeling Language (SysML)
System, 24, 25
Systematic engineering, 30
Systematic program, 26
System life cycle, 63–68, 235
System of interest, 116, 117
System of systems (SoS), 27, 42, 57
    activities, 32
    aircraft, maintenance, and logistics, 44
    behavior, 30
    coherent architecture, 46
    constituent parts, 46
    elementary subprocesses, 47, 48
    elements, 30
    impact, 33
    real environment, 33
    stakeholders, 33

Systems engineering, 158, 235
Systems Modeling Language (SysML), 65–67, 78, 80

T

Tethered XR devices, 13
Theory of Constraints (TOC), 97
Three-dimensional (3D) form, 73
Three-dimensional (3D) model, 36
Time synchronization, 169
   importance, 155
   protocol, 169
   server, 170
TOC, *see* Theory of Constraints (TOC)
Total Quality Management (TQM), 97
TQM, *see* Total Quality Management (TQM)
Training, 130–133
Troy horse, 183
Tutorials, 131

U

Uncertainty, 22–24, 33–37, 58, 59, 235

V

Validation, 64, 69, 71–76, 78, 79, 90, 101–103, 121, 236
   design specifications, 101

Value-added activities, 68
VE, *see* Virtual environment (VE)
Verification, 69–71, 78, 79, 236
Verification and validation (V&V), 70
Virtual engineer, 90
Virtual environment (VE), 44, 64
Virtual Reality (VR), 1, 6, 9, 19, 36, 56, 58, 61, 75, 78, 96, 177, 219, 236
   advantages, 150
   applications, 126–129
   headsets, 136
   industrial applications, 192
Visualization, 65, 66, 72, 73, 77, 189, 191
VR, *see* Virtual Reality (VR)
Vu, Francis, 200–202
V&V, *see* Verification and validation (V&V)

W

Wheeler, Ryan, 195–197

X, Y

XR deployment
   applications, 172
   demonstration sessions, 179, 180
   industrial environment, 175, 176
   plan development, 173

INDEX

XR deployment (*cont.*)
   reinvention, 176
   systematic approach, 173
   transformative part, 180, 181
   unions and work councils,
      185, 186
XR gadgetry, 162
XR4MBSE, 83, 84
XR sickness, 14
XR technology, 2–4, 22, 23, 54–56
   advantages, 150
   AI issues, 15, 16
   challenges, 194
   disciplines, 88, 89
   factors, 84, 85
   industrial metaverse, 157
   MBSE integration, 193
   products and services, 128
   remote assistance, 129, 130
   system life cycle, 63–65
   training, 130–133

   verification, 69–71
   VR applications, 126–129
XR training, 130–133, 138
   AIShed survey, 133–138
   assembly, 140, 141
   collaboration, 142–144
   guidance, 139, 140
   navigation, 145–148

Z

ZapWorks, 3
ZDM, *see* Zero defect
      manufacturing (ZDM)
Zero defect manufacturing
   (ZDM), 120
Zero defect (ZD) program
   automatic simulations,
      100, 101
   group beta, 101–106
   instrumentation, 97

Printed in the United States
by Baker & Taylor Publisher Services